*The Mitchell Beazley pocket guide to*
# Mushrooms and Toadstools

David N. Pegler

Mitchell Beazley

## Nomenclature

Unlike most plants, mushrooms and toadstools do not always have common or English names, probably because of the traditional fear many people seem to share of these strange, fleshy organisms. All have Latin or scientific names, however, and while they may be initially off-putting to the beginner, they are the only reliable means of correctly labelling a fungus. These Latin names consist of two parts: the first gives the *genus* name (e.g. *Amanita*), and the second gives the *species* name within that genus (e.g. *muscaria*, *fulva*, *phalloides*). A genus (plural genera) is a collective term used to describe a group of species with certain features in common, and is frequently abbreviated to the first letter when there is reference to more than one species within that group. Hence we have *Amanita muscaria*, *A. fulva* and *A. phalloides*. In some instances, especially where the genus is very large, species may be grouped into subgenera, but this does not manifest itself in the Latin name.

The increasing use of the microscope has revealed fundamental differences at the cellular level between species which were formerly placed in the same genus, and consequently many species have been renamed. All the common names given in this book are those currently approved by the British Mycological Society, and where possible any names by which a species was previously known are included in the text.

## Abbreviations

| | | | | | |
|---|---|---|---|---|---|
| cm | centimetre | g | gramme | sp(p) | species (plural) |
| diam | diameter | min | minute | μm | micron (0.001 mm) |
| esp | especially | mm | millimetre | | |

## Symbols

**Edibility**—see also p 6.

- ⓧ Edible and excellent
- ⓦ Edible only after cooking
- ⓧ Inedible, unpleasant or indigestible
- ⓟ Poisonous
- ⓓ Deadly poisonous

**Frequency**

- Ⓒ Common
- Ⓕ Fairly common
- Ⓞ Occasional
- Ⓡ Rare in Britain

**Habitat**—the following symbols are only used to indicate other environments in which a species may be found (or to reinforce preference for a particular tree such as beech in the "Beech and oak woods" section).

- 🌲 Coniferous woods
- 🌳 Oak woods
- 🌳 Beech woods
- 🌳 Frondose woods
- 🌳 Birch woods
- 🌾 Grassland
- 🌿 Marshland

---

**Editor** Michele Staple **Designers** Hazel West and Jacquie Gulliver
**Illustrators** Patrick Cox / The Garden Studio; Gill Tomblin; Colin Emberson

Edited and designed by Mitchell Beazley Publishers
part of Reed International Books Ltd, Michelin House,
81 Fulham Road, London SW3 6RB

© Mitchell Beazley Publishers 1981
reprinted 1982, 1984, 1987, 1988, 1990, 1991, 1992
All right reserved
ISBN 0 85533 347 2

Colour reproduction by Lithospeed Ltd
Printed in Malaysia

## Contents

Mushroom structure 4
Glossary 4
Collecting mushrooms 6
Edible mushrooms 6
Poisonous mushrooms 6
Field key 7
Coniferous woods 12
Mixed woods 43
Frondose woods 62
Birch woods 95
Beech and oak woods 103
Alder carrs 125
Elm 127
Parks, roadsides and wasteland 129
Heaths and grassland 141
Marshes, fens and bogs 154
Sand dunes and seaside meadows 158
Burnt ground 160
Dung 162
Index 166

## Introduction

"Mushroom" and "toadstool" are terms rather loosely applied to the fruitbodies of fleshy gill-fungi, and are commonly (if somewhat inaccurately) used to denote edible and poisonous species respectively. They form but a small part of the enormous range of organisms known as fungi, amongst which are included the puffballs, moulds, mildews, cup fungi, rusts, smuts and yeasts. Their essential characteristic is the lack of the green pigment chlorophyll, and this puts the fungi in a separate kingdom from plants. Plants are *producers*: using chlorophyll and energy from sunlight, they manufacture and store their own food by a process called photosynthesis. Fungi are not producers, like plants, nor are they *consumers*, like animals, which ingest their food. Instead they secrete enzymes with which they break down substances in order to obtain food—they are the *reducers*.

Fungi are made of filaments, called hyphae, which branch out to form an extensive web (the *mycelium*). This is often hidden underground, so that the part most people notice is usually the fruitbody. The fruitbody is the reproductive part of the fungus and arises from the mycelium to release masses of tiny spores (invisible to the naked eye) into the air stream.

Fungi fall into three groups according to the way they produce spores:

1 The "Phycomycetes"—these are entirely microscopic fungi which do not form complex fruitbodies but have simple hyphae which grow upright to form a *sporangium* (e.g. bread mould).

2 Ascomycotina—these produce spores *inside* elongated terminal cells; each cell is called an *ascus* (e.g. cup fungi).

3 Basidiomycotina—here the spores are borne on stalks *outside* the terminal cells; each cell is called a *basidium* (e.g. mushrooms and toadstools).

## How to use this book

Mushrooms and toadstools of the same genus do not necessarily grow in the same habitat, and it is for this reason that the species in this book have been grouped according to their most characteristic habitat. While this will prove invaluable in the field, you must remember that a number of species will occur in habitats additional to those in which they are illustrated. If you cannot locate a species by looking in the relevant section, then turn to the field key on p 7. Symbols will also inform you of alternative habitats, as well as giving guidance on frequency and edibility.

Descriptions for each species include cap diameter, shape, colour range, gill attachment, stem height and season (in italics). It is a good idea to learn the basic structure of a mushroom (p 4) and to familiarize yourself with some of the technical terms explained in the glossary (p 4) before venturing out into the field. The section drawings, apart from showing cap shape and gill attachment, also indicate flesh colour (and if it changes on exposure to air). Advanced students will find the spore illustrations and details particularly useful if they have access to a microscope.

# Mushroom structure

*In the boletes, the gills are replaced by vertical tubes*

The mushroom is like an umbrella, with a cap protecting the spore-producing surfaces (the *gills*) from the rain. The young fruitbody or *primordium* is initially enclosed by one or two *veils*, and when conditions are favourable, it expands by absorbing water, rupturing the outer (or *universal*) veil and allowing the stem to lengthen. The inner (or *partial*) veil protects the young gills, and as the cap expands, this also breaks. The inner veil may persist as a ring on the stem, while the outer veil may leave a volva at the base and/or scales on the cap.

A mushroom begins its life when a spore lands on a substratum (often wood or soil) suitable for germination to take place. Hyphae radiate from the spore to produce a mycelium, which may be one of two or more different mating types. When two hyphae from compatible mycelia of the same species meet, they fuse together and produce a bud-like formation (the primordium). It is from this primordium that the characteristic mushroom shape develops.

The main function of a mushroom is to produce spores, and it does so by means of special club-shaped cells called basidia (see illustration on p 5). These basidia cover the surfaces of the gills (tubes in the case of boletes), and bear spores on protruding stalks. Each basidium typically releases four spores, which are dispersed by air currents. The distinctive coloration of the spores accounts for the change in gill colour as the fruitbody matures.

The structure of a fruitbody is important in the classification of mushrooms or agarics (order Agaricales). The way in which the gills are attached to the stem often characterizes a genus and some species are recognized by the shape of the cap. Modern classification, however, places more emphasis on microscopic features and chemical tests, with spore structure proving to be the most important characteristic of all. Today 17 families are recognized in the Agaricales, and in Britain alone there are over 1,850 mushroom species.

## Glossary

bulbous

cortina

interveining

**Adnate** (of gills, tubes) broadly attached to stem
**Adnexed** (of gills, tubes) narrowly attached to stem
**Agaric** one of the gill-bearing fungi; mushrooms
**Amyloid** (of spores) staining dark blue in Melzer's reagent (solution containing iodine)
**Appressed** (of scales, hairs) flattened down
**Arcuate-decurrent** (of gills, tubes) distinctly curved and descending down the stem
**Ascending** (of gills, tubes) slanting upwards to almost vertical
**Basidium** a club-shaped cell producing spores (usually 4) on outer stalks
**Bolete** a quickly rotting fungus, characterized by tubes and pores underneath the cap, instead of gills
**Bulbous** (of stem) enlarged at base
**Campanulate** (of cap) bell-shaped
**Cap** upper part of mushroom which bears gills
**Cartilaginous** hard and tough, breaking with a snap
**Chlamydospores** thick-walled, asexual spores
**Compressed** laterally flattened
**Conical** (of cap) shaped like a cone
**Convex** (of cap) broadly curved or rounded

**Cortina** a cobweb-like veil, initially joining cap margin to stem
**Cyathiform** (of cap) deeply depressed (or funnel-shaped with a broad base)
**Decurrent** (of gills, tubes) broadly attached and descending down the stem
**Deliquescent** dissolving into a liquid
**Dextrinoid** (of spores) staining reddish brown in Melzer's reagent (solution containing iodine)
**Ellipso-cylindric** (of spores) short, rod-shaped
**Emetic** causes vomiting
**Ephemeral** short-lived, soon disappearing
**Excentric** off-centre
**Expanded** (of cap) opening out when mature
**Fibrillose** covered with small fibres
**Fibrils** small fibres
**Free** (of gills, tubes) not attached to stem
**Frondose** (of trees) deciduous or broad-leafed
**Fruitbody** the entire mushroom or toadstool (excluding the mycelium)
**Germ-pore** (of spores) minute opening in the spore wall through which the spore germinates
**Gills** radiating structures underneath the cap on which the spores are produced
**Glutinous** very slimy
**Haemolysins** toxins capable of destroying red blood cells
**Hypha(-e)** individual filaments which collectively form the mycelium and fruitbody
**Infundibuliform** (of cap) funnel-shaped
**Interveining** (of gills) ridges connecting adjacent gills or partially so
**Lateral** (of stem) attached to side of cap
**Milk** liquid released by flesh of fruitbody when broken (frequently species of *Lactarius*, the milk caps)
**Mycelium** a mass of hyphae
**Mycorrhizal association** symbiotic relationship between a fungus and the roots of a tree
**Ovoid** more or less egg-shaped
**Ovoid-angular** (of spores) egg-shaped with flattened sides
**Papilla** nipple, small projection
**Partial veil** protective layer of immature fruitbody, covering the young gills
**Perisporium** loose, outermost layer of spore wall
**Polygonal** (of spores) angular, with many flat sides
**Pores** the orifices of the tubes in the boletes
**Radially fibrillose** fibres radiating outwards
**Recurved** (of scales) curved upwards
**Rhizomorph** cord-like strand of mycelium, extending over or through the substratum
**Ring** membrane encircling stem, derived from veil
**Rosaceous** of the rose family
**Sclerotium** small, sterile mass of closely packed hyphae
**Sinuate** (of gills) curving abruptly upwards near point of attachment to stem
**Spindle-shaped** tapering at both ends
**Spores** reproductive cells, capable of germinating and producing another fungus
**Stem** lower part of fruitbody supporting the cap
**Striate** lined or grooved
**Substratum** material on which the fungus grows
**Triangular** (of gills) broadly decurrent
**Tubes** spore-producing surfaces arranged vertically underneath the cap in a dense layer; found in boletes
**Umbilicate** (of cap) with a small central depression
**Umbonate** (of cap) with a central hump (umbo)
**Universal veil** outermost protective layer covering immature fruitbody
**Veil** protective layers of immature fruitbody
**Volva** remains of veil, forming a cup-like structure at stem base

convex
free
expanded
adnexed
flattened
sinuate
umbonate
adnate
depressed
decurrent
umbilicate
triangular
infundibuliform
arcuate-decurrent
conical
ascending
campanulate
ascending with a decurrent tooth
basidium

## Collecting mushrooms

If you have never collected mushrooms before, then try to attend one of the fungus forays arranged in the autumn by your local natural history society or the British Mycological Society for expert advice. Before setting out, make sure you have a flat-bottomed basket, a selection of small tins or waxed paper for keeping different mushrooms apart, and a note pad. It is always advisable to make notes in the field, particularly recording the locality and substratum, variations in colour and any distinctive smell.

Pick your mushroom carefully, gently lifting the entire fruitbody, including the base, so that you do not damage the delicate structures. Examine the fruitbody closely, noting its size, colour, shape and texture. The colour of the spores does not always agree with the colour of the gills, so make a spore print by placing the cap with the gills downwards on a sheet of white paper and leaving it overnight under a glass jar.

## Edible mushrooms

The golden rule when picking mushrooms to eat is *never* to experiment with fungi which you have not identified with certainty as harmless. Rule number two is to disregard totally any folk lore giving "tests" on safety, for they are known to be unreliable on certain lethal species. Having said that, it is only fair to point out that there are many fungi which are both edible and excellent, and in many countries they have formed part of the human diet for centuries. In eastern Europe they are highly valued as a food source, and over 30 species are sold commercially. Edible mushrooms contain more protein than any vegetable, and are rich in vitamin B.

The most delicious mushrooms include the Penny bun boletus (p 123), the Miller (p 149), the Parasol mushroom (p 146) and the Chanterelle (p 103), to name but a few. All edible fungi should be picked when young and fresh, and cooked as soon as possible (after removing any tough or slimy parts).

## Poisonous mushrooms

Serious cases of poisoning occur every year, most of which are caused by less than ten species. Fortunately most of these are rare in Britain, but it is still important to be able to recognize them and to know something of their various effects. Beginners should avoid eating species of *Amanita*, *Cortinarius*, *Inocybe*, *Clitocybe*, *Tricholoma*, *Lepiota* and the peppery species of *Russula* and *Lactarius*, and keep to fresh mushrooms which have been accurately identified. If you do suspect poisoning, *then seek medical help immediately*, for delay can prevent cure. Wherever possible, supply the doctor with a sample of the suspected mushroom, even if only a fragment.

Clinically the forms of mushroom poisoning fall into eight categories:

**1 Cytolytic poisoning.** *(a) Cyclopeptides*—by far the most dangerous toxins, attacking the kidneys and liver although no symptoms may be apparent for at least six hours. *Amanita phalloides* and *A. virosa* (p 104) are responsible for nearly every death by mushroom poisoning. Several *Lepiota* species and *Galerina unicolor* (p 28) also contain cyclopeptides. *(b) "Orellanin"*—found in *Cortinarius speciosissimus* (p 30), causes fatal damage to kidneys; symptoms may not be felt until after 3–17 days.

**2 Haemolysins.** These attack blood cells, but their activity is destroyed by cooking. Such mushrooms include *Amanita vaginata* and *A. fulva* (p 96).

**3 Muscarine poisoning.** Causes perspiration and nausea after two hours, but is not generally fatal. Species include *Amanita muscaria* (p 95), *A. pantherina* (p 44) and *Clitocybe dealbata* (p 143).

**4 Psychotropic poisoning.** *(a) Ibotenic acid and muscimol groups*—these cause hallucination and/or delirium, leading to torpidity and possible coma; e.g. *Amanita muscaria* (p 95). *(b) Indole groups, containing psilocin and psilocybin*—hallucinogenic toxins, stimulating psychic perception; e.g. *Psilocybe semilanceata* (p 138). Possession and consumption of psilocin for "recreational purposes" are offences under the Misuse of Drugs Act, 1970.

**5 Coprine poisoning.** When consumed with alcohol, *Coprinus atramentarius* (p 139) can be extremely unpleasant, causing vomiting and sickness.

**6 Gastro-enteric irritants.** Many mushrooms fall into this group—their effect depends on the amount consumed and the individual's susceptibility. Species include *Entoloma sinuatum* (p 133), *Agaricus xanthodermus* (p 136) and many species of *Tricholoma*, *Russula* and *Lactarius*.

**7 Intolerance and allergies.** Only some people are affected. Species include *Clitocybe nebularis* (p 13) and *Armillaria mellea* (p 45).

**8 Gastro-enteritis.** Results from eating mouldy mushrooms.

# Field key

This key has been designed for use in the field, and is simple to use, provided you keep to a few basic rules. Only those species illustrated in the book are covered by the key, which should guide you to the correct genus. You can then identify the species by referring to the illustrations and text. Page numbers are given with each genus wherever it occurs in the book, which will help you to locate the species more quickly.

When you have selected your specimen, work through the key, choosing between the two options (**a** and **b**) after careful observation of such features as gill attachment, shape, colour and texture, as well as habitat. You will not need a microscope for this key to determine spore structure, but you will need to know the colour of the spores, and this can be discovered by making a spore print (see Collecting mushrooms, p 6). Spore colours vary in intensity and may be tinged with other colours, but broadly speaking they fall into five categories:

☐ white  ☐ pink  ▨ yellowish brown  ▰ dark brown  ■ black

### Preliminary key

| | | Proceed to |
|---|---|---|
| 1a | Cap with pores or tubes underneath | **Key A** (below) |
| 1b | Cap with gills (or ridges) underneath | 2 |
| 2a | Stem excentrically to laterally attached to cap, or absent | **Key B** (below) |
| 2b | Stem centrally attached to cap | 3 |
| 3a | Gills free or narrowly attached | 4 |
| 3b | Gills broadly adnate to decurrent | 5 |
| 4a | Gills free | **Key C** (p 8) |
| 4b | Gills narrowly attached, adnexed or shortly adnate | **Key D** (p 8) |
| 5a | Gills broadly adnate or sinuate | **Key E** (p 9) |
| 5b | Gills decurrent | **Key F** (p 10) |

### Key A Cap with pores and tubes underneath; fruitbody soft, quickly rotting

| | | |
|---|---|---|
| 1a | Spore print dark brown to blackish; cap with thick, dark scales | *Strobilomyces floccopus* 94 |
| 1b | Spore print yellowish brown or pinkish | 2 |
| 2a | Stem very swollen, up to 5–15 cm diam, with network pattern on upper surface | *Boletus* 60, 121–123 |
| 2b | Stem elongate, cylindric, rarely exceeding 4 cm diam | 3 |
| 3a | Spore print (and pores) pink; tubes long; cap dry | *Tylopilus felleus* 124 |
| 3b | Spore print yellowish brown to cinnamon | 4 |
| 4a | Stem hollow; tubes free, 3–6 mm long | *Gyroporus* 124 |
| 4b | Stem solid | 5 |
| 5a | Stem with minute, glandular scales or a membranous ring | 6 |
| 5b | Stem not scaly, often slender | 7 |
| 6a | Cap sticky or slimy; always associated with coniferous trees | *Suillus* 40–41 |
| 6b | Cap dry; stem very long, ridged, with numerous darker scales | *Leccinum* 102 |
| 7a | Tubes very decurrent, short; pores large radially arranged; under alder | *Gyrodon lividus* 126 |
| 7b | Tubes adnate; pores small to medium, yellow or red, often bruising bluish | *Xerocomus* 39, 61 |

### Key B Cap with gills; stem excentrically to laterally attached, or absent

| | | |
|---|---|---|
| 1a | Spore print white or pink | 2 |
| 1b | Spore print yellowish brown to dark brown; fruitbody small | 10 |
| 2a | Fruitbody tough, leathery, dry | 3 |
| 2b | Fruitbody soft or gelatinous | 5 |
| 3a | Cap small, 1–3 cm diam, greyish, lacks a stem; gills divide along their edge | *Schizophyllum commune* 110 |
| 3b | Cap brown, with a stem | 4 |

| | | |
|---|---|---|
| 4a | Cap fan-shaped or irregular; stem short | **Panus** 71 |
| 4b | Cap funnel-shaped; stem well developed | **Lentinellus cochleatus** 79 |
| 5a | Fruitbody with jelly-like layer | 6 |
| 5b | No gelatinized layer present | 9 |
| 6a | Fruitbody with a stem | 7 |
| 6b | Fruitbody without stem, laterally attached to substratum | 8 |
| 7a | Stem excentric; cap orange with wrinkled pellicle; gills sinuate | **Rhodotus palmatus** 128 |
| 7b | Stem lateral, short; cap white or olive; gills adnate | **Panellus** 21, 109 |
| 8a | Cap very small, up to 1 cm diam, dark grey; gills grey | **Resupinatus applicatus** 69 |
| 8b | Cap up to 8 cm diam, spoon-shaped; gills white | **Hohenbuehelia petaloides** 79 |
| 9a | Cap without stem, laterally attached, with bright orange gills | **Phyllotopsis nidulans** 21 |
| 9b | Cap without stem (or very short), fan-shaped or funnel-shaped, with white gills | **Pleurotus** 71, 110 |
| 10a | Fruitbody without stem, laterally attached, on wood; gills soon yellowish brown | **Crepidotus** 82 |
| 10b | Cap up to 8 mm diam, with a short, lateral stem; on *Carex* and grass debris; gills finally dark brown | **Melanotus phillipsii** 153 |

# Key C Cap with free gills; stem central

| | | |
|---|---|---|
| 1a | Volva present at stem base, typically as membranous sac, but sometimes just scales | 2 |
| 1b | Volva absent | 3 |
| 2a | Spore print pink; no ring on stem | **Volvariella** 82, 128 |
| 2b | Spore print white; stem with or without a ring | **Amanita** 37, 43–44, 62, 95–96, 104 |
| 3a | Ring or granular scales present on stem; spore print white, dark brown or black | 4 |
| 3b | Stem lacking a ring; spore print pink or yellowish brown | 10 |
| 4a | Spore print dark brown or black | 5 |
| 4b | Spore print whitish | 6 |
| 5a | Large, fleshy fruitbodies, often whitish, usually with ring on stem | **Agaricus** 35–37, 59, 91, 136–137, 152–153, 159 |
| 5b | Small, greyish-black fruitbodies, with granular scales on cap and stem | **Melanophyllum echinatum** 74 |
| 6a | Cap glutinous, smooth, pale ochre; stem tall | **Limacella guttata** 72 |
| 6b | Cap dry | 7 |
| 7a | Large, fleshy fruitbodies; stem more than 5 mm diam | 8 |
| 7b | Fruitbodies more slender, unless cap covered with pyramidal scales | **Lepiota** 50, 73–74, 133 |
| 8a | Cap pale, smooth, stem with simple, membranous ring | 9 |
| 8b | Cap scaly; stem with complex double ring | **Macrolepiota** 133, 146–147 |
| 9a | Stem with minute, yellowish-brown scales; gills narrow | **Chamaemyces fracidus** 72 |
| 9b | Stem smooth; gills broad | **Leucoagaricus naucinus** 147 |
| 10a | Spore print pink | **Pluteus** 81 |
| 10b | Spore print yellowish brown; cap and stem yellow, fragile; on dung | **Bolbitius vitellinus** 162 |

# Key D Cap with adnexed to shortly adnate gills; stem central

| | | |
|---|---|---|
| 1a | Spore print whitish, rarely with a pale pinkish tint | 2 |
| 1b | Spore print deep pink, yellowish brown or black | 12 |
| 2a | Large, dull-coloured or whitish fruitbodies; cap 3–8 cm diam or more; stem more than 5 mm diam | 3 |

| | | |
|---|---|---|
| **2b** | Fruitbody smaller or brightly coloured | **4** |
| **3a** | Cap glutinous; on dead branches or with a rooting stem | ***Oudemansiella* 106** |
| **3b** | Cap dry; mostly on leaf litter in woods | ***Collybia* 15, 46–47, 66, 105–106** |
| **4a** | Cap conical to campanulate; gills ascending | **5** |
| **4b** | Cap convex or flattened | **7** |
| **5a** | Fruitbody waxy, red, orange or yellow; found on grassland | ***Hygrocybe* 144–146** |
| **5b** | Fruitbody not waxy; stem long, slender | **6** |
| **6a** | Fruitbody dark brown, smell of fish or cucumber | ***Macrocystidia cucumis* 20** |
| **6b** | Fruitbody paler or with bright colours; smell nitrous or none | ***Mycena* 18, 50, 67–68, 107–108, 125, 148** |
| **7a** | Very small; in specialized habitats | **8** |
| **7b** | Larger species | **9** |
| **8a** | On fallen and buried cones | ***Baeospora myosura* 19; *Strobilurus esculentus* 19** |
| **8b** | On rotting fungi | ***Nyctalis* 47, *Collybia tuberosa* 47, *C. cirrhata* 67** |
| **9a** | Fruitbody tough, stem wiry | **10** |
| **9b** | Fruitbody soft, fleshy, easily decaying | **11** |
| **10a** | Cap smooth; on twigs and dead leaves | ***Marasmius* 18, 69, 109, 131, 147; *Marasmiellus ramealis* 69** |
| **10b** | Cap and stem hairy; on dead grass stems | ***Crinipellis stipitaria* 149** |
| **11a** | Cap yellow, orange, or pink, powdery; stem with a ring | ***Cystoderma* 22** |
| **11b** | Cap dark brown, smooth, striate; no ring | ***Tephrocybe* 158, 160** |
| **12a** | Spore print pink | **13** |
| **12b** | Spore print yellowish brown or black | **14** |
| **13a** | Cap brown, smooth, striate | ***Nolanea* 27, 150–151** |
| **13b** | Cap bluish, purplish, yellow and/or scaly | ***Leptonia* 149–150** |
| **14a** | Spore print yellowish brown | **15** |
| **14b** | Spore print black (or blackish brown) | **20** |
| **15a** | Fruitbody large and fleshy | **16** |
| **15b** | Cap thin-fleshed; stem long, slender | **18** |
| **16a** | Stem with a membranous ring | **17** |
| **16b** | No ring; cap conical, radially fibrillose | ***Inocybe* 57, 84–86, 119, 135, 159** |
| **17a** | Cap golden brown, granular | ***Phaeolepiota aurea* 88** |
| **17b** | Cap dull ochre, not granular | ***Rozites caperata* 34** |
| **18a** | Cap glutinous, with pointed umbo; stem rooting | ***Phaeocollybia festiva* 28** |
| **18b** | Cap dry or moist but not sticky | **19** |
| **19a** | In damp places, associated with moss and alder carrs | ***Naucoria* 126; *Galerina* 28, 86, 156** |
| **19b** | Amongst grass, sand or dung | ***Conocybe* 135, 158, 162–163** |
| **20a** | Gills crowded, deliquescent | ***Coprinus* 91–92, 120, 139–140, 163–164** |
| **20b** | Gills quite spaced, not deliquescent | ***Psathyrella* 34, 92–94, 120, 139, 159, 161** |

**Key E** Cap with broadly adnate or sinuate gills; stem central

| | | |
|---|---|---|
| **1a** | Spore print whitish or very pale pink | **2** |
| **1b** | Spore print deeply coloured, pink, yellowish brown, dark brown or black | **13** |
| **2a** | Fruitbody mostly large and fleshy, with stem 5 mm or more thick | **3** |
| **2b** | Fruitbody more slender, stem less than 5 mm thick | **9** |
| **3a** | Cap and stem brittle, not fibrous; cap often brightly coloured | ***Russula* 26–27, 51–53, 75–76, 99, 112–114** |
| **3b** | Cap and stem fleshy or fibrous | **4** |
| **4a** | Stem tall, striate; cap smooth, greyish brown or pale | ***Melanoleuca* 75, 130, 148** |
| **4b** | Stem typically shorter, or cap scaly | **5** |
| **5a** | Growing on wood, or from a conspicuous, white, cord-like mycelium | ***Tricholomopsis* 17, 48** |

| | |
|---|---|
| 5b Growing on soil | 6 |
| 6a Tufted; cap smooth, white or dull-coloured | *Lyophyllum* 21, 127, 129–130 |
| 6b Generally solitary | 7 |
| 7a Cap, gills and/or stem with lilac or mauve colours | *Lepista* 51, 65, 132, 148 |
| 7b Gills never lilac or mauve | 8 |
| 8a Woodland species; cap scaly or fibrillose | *Tricholoma* 15–16, 48, 70, 97, 111, 132 |
| 8b Grassland species; cap cream or pinkish, not scaly | *Calocybe* 141–142 |
| 9a Cap conical or convex, not expanding | 10 |
| 9b Cap convex expanding to depressed | 12 |
| 10a Fruitbody waxy, orange, red or scarlet | *Hygrocybe* 144–146 |
| 10b Fruitbody fibrous or fleshy | 11 |
| 11a Gills ascending | *Mycena* 18, 50, 67–68, 107–108, 125, 148 |
| 11b Gills horizontal; cap dark brown | *Fayodia bisporigera* 49 |
| 12a Growing on soil; fruitbody pinkish brown or mauve | *Laccaria* 66, 154 |
| 12b Tufted on wood; cap yellowish, sticky; stem brown, velvety | *Flammulina velutipes* 67 |
| 13a Spore print and gills deep pink; gills sinuate | *Entoloma* 27, 80, 133, 149 |
| 13b Spore print yellowish brown, dark brown or black | 14 |
| 14a Spore print yellowish brown | 15 |
| 14b Spore print dark brown or blackish | 21 |
| 15a Growing typically on wood | 16 |
| 15b Growing on soil | 18 |
| 16a Stem scaly, with a ring | 17 |
| 16b Stem either without a ring or lacking scales; gills bright rust brown | *Gymnopilus* 28, 88 |
| 17a Cap smooth, date brown when moist, drying paler | *Kuehneromyces mutabilis* 89 |
| 17b Cap dry, usually scaly, yellow or brown | *Pholiota* 29, 58, 89–90, 121, 126, 157, 161 |
| 18a Small species, with conical cap and long, slender stem; with mosses | *Galerina* 28, 86, 156 |
| 18b Larger, fleshier fruitbodies | 19 |
| 19a Stem with membranous ring; cap white or dull brown | *Agrocybe* 89, 134, 151 |
| 19b Typically without a membranous ring but often a cobweb cortina in young stage | 20 |
| 20a Cap smooth, sticky; fruitbody pale brown | *Hebeloma* 28, 86–87, 119, 161 |
| 20b Cap often scaly, dry or slimy; fruitbody often brightly coloured | *Cortinarius* 29–32, 56–57, 83, 100–101, 116–118, 151, 155; *Dermocybe* 33, 155 |
| 21a Fruitbody slender; stem with a ring or cobweb-like cortina | 22 |
| 21b Fruitbody slender; stem without ring | 24 |
| 22a Gills mottled; stem with conspicuous fibrillose veil | *Lacrymaria velutina* 139 |
| 22b Gills not mottled | 23 |
| 23a Fruitbody solitary, on soil or dung | *Stropharia* 38, 91, 152, 165 |
| 23b Fruitbody often tufted, usually on wood | *Hypholoma* 38, 151, 157 |
| 24a Gills mottled, ascending | 25 |
| 24b Gills not mottled, horizontal; fruitbodies small, on dung or in grass | *Psilocybe* 120, 138, 153, 156, 165 |
| 25a Stem long; on dung or in grass | *Panaeolus* 138, 164 |
| 25b Stem shorter; on lawns | *Panaeolina foenisecii* 138 |

## Key F  Cap with decurrent gills; stem central

| | |
|---|---|
| 1a Cap funnel-shaped, brownish grey, without true gills or ridges | *Craterellus cornucopioides* 105 |
| 1b Cap with well developed gills or ridges | 2 |
| 2a Gills with conspicuous forking | 3 |

| | | |
|---|---|---|
| **2b** | Gills with little or no forking | **5** |
| **3a** | Gills thin, white; cap and stem ash grey | ***Cantharellula umbonata*** 143 |
| **3b** | Gills bright yellow or orange | **4** |
| **4a** | Gills thick with blunt edges | ***Cantharellus*** 66, 103, 105 |
| **4b** | Gills thin; in coniferous woods | ***Hygrophoropsis aurantiaca*** 17 |
| **5a** | Spore print whitish | **6** |
| **5b** | Spore print pink, yellowish brown or very dark brown (blackish brown) | **21** |
| **6a** | Fleshy or tough fruitbodies; cap up to 5 cm diam, stem 5 mm or more thick | **7** |
| **6b** | Fruitbody small, not fleshy; cap umbilicate | **16** |
| **7a** | Fruitbody brittle, releasing a white or coloured milk when broken | ***Lactarius*** 23–24, 54–55, 77–78, 97–98, 114–115, 125 |
| **7b** | No milk present | **8** |
| **8a** | Fruitbody waxy | **9** |
| **8b** | Fruitbody not waxy | **11** |
| **9a** | Brightly coloured, yellow, red or orange | ***Hygrocybe*** 144–146 |
| **9b** | Dull-coloured or white | **10** |
| **10a** | Cap sticky | ***Hygrophorus*** 13, 63, 108 |
| **10b** | Cap dry | ***Camarophyllus*** 142–143 |
| **11a** | Fruitbody tough, persistent, white | **12** |
| **11b** | Fruitbody soft, fleshy | **13** |
| **12a** | Cap scaly; gill edge often toothed; on wood | ***Lentinus*** 20 |
| **12b** | Cap smooth; fruitbody tufted; on soil | ***Lyophyllum connatum*** 129 |
| **13a** | Stem with a ring; cap scaly | ***Armillaria*** 45 |
| **13b** | Stem without a ring | **14** |
| **14a** | Fruitbody very large, whitish; cap infundibuliform | ***Leucopaxillus giganteus*** 131 |
| **14b** | Fruitbody either small or not infundibuliform | **15** |
| **15a** | Cap, gills and flesh greyish brown; stem tall | ***Pseudoclitocybe*** 65 |
| **15b** | Cap and gills paler or differently coloured; stem shorter | ***Clitocybe*** 13–14, 45–46, 63–64, 143 |
| **16a** | Growing on wood | **17** |
| **16b** | Growing on soil or amongst mosses | **18** |
| **17a** | Cap rusty yellow or ochre; stem smooth | ***Xeromphalina*** 19, 49 |
| **17b** | Cap dark reddish brown; stem velvety | ***Micromphale*** 18, 49 |
| **18a** | Cap whitish and margin fringed with short hairs; in coniferous woods | ***Ripartites tricholoma*** 19 |
| **18b** | No fringe on cap margin | **19** |
| **19a** | On burnt ground; cap greyish brown to black | ***Myxomphalia maura*** 161 |
| **19b** | Not associated with burnt ground | **20** |
| **20a** | Amongst moss; cap yellowish orange | ***Gerronema*** 68, 147 |
| **20b** | Amongst short grass or in damp places; cap white or brownish | ***Omphalina*** 147, 155, 158 |
| **21a** | Spore print pink | **22** |
| **21b** | Spore print yellowish brown or very dark brown (blackish brown) | **24** |
| **22a** | Fruitbody white | ***Clitopilus*** 82, 149 |
| **22b** | Fruitbody greyish brown or pinkish brown | **23** |
| **23a** | Cap fleshy | ***Rhodocybe*** 20, 159 |
| **23b** | Cap membranous, umbilicate | ***Eccilia sericeonitida*** 154 |
| **24a** | Spore print yellowish brown (clay or coffee) | **25** |
| **24b** | Spore print very dark brown (blackish brown); cap sticky; in coniferous woods | **27** |
| **25a** | Cap fleshy | **26** |
| **25b** | Cap thin, scurfy scaly | ***Tubaria furfuracea*** 135 |
| **26a** | Gills crowded, brown, slimy; cap margin inrolled | ***Paxillus*** 39, 101 |
| **26b** | Gills widely spaced, interveined, thick, yellow | ***Phylloporus rhodoxanthus*** 116 |
| **27a** | Young gills covered by a glutinous veil | ***Gomphidius*** 42 |
| **27b** | No glutinous veil | ***Chroogomphus rutilus*** 42 |

# Coniferous woods

Coniferous trees are cone-bearing, needle-leaved or scale-leaved, and mostly evergreen. They include the pines, spruces, firs, larches and cypresses. The needles form a thick carpet on the forest floor, which eventually breaks down into a dark, compact humus layer called mor. The soil is rather poor and the nitrogen, mineral and organic content is very low so that the only plants covering the forest floor tend to be mosses, liverworts and some ferns. Mushrooms and toadstools, however, can occur in very large numbers, often growing in troops or circles. In many cases a symbiotic relationship exists in which the underground mycelial threads of the fungus are intimately connected with the roots of the tree. Neither the fungus nor the tree can grow without the other, and both gain nutritionally from this relationship, which is known as an *ectotrophic mycorrhizal association*. Many mushrooms are specifically associated with one host tree species, although a tree may grow with a number of different mushroom species. This is one of the main reasons why certain mushroom and toadstool species are only to be found in coniferous woods.

Amongst the larger mycorrhizal mushrooms are the milk caps (*Lactarius*), which release a milky fluid when broken, the related *Russula* species, which are frequently reddish or tinted violet, and innumerable *Cortinarius* species, which have a cobweb-like veil covering the young gills. Several of the boletes are also restricted to conifers, especially those of the genus *Suillus*; similarly the spike-caps (*Chroogomphus*, *Gomphidius*) prefer this habitat. On fallen twigs and needles occur several very small fungi such as *Marasmius androsaceus*, *Mycena sanguinolenta* and *Micromphale perforans* (p 18), whilst *Baeospora myosura* and *Strobilurus esculentus* (p 19) actually grow from the fallen cones. The stumps of coniferous trees may produce conspicuous mushrooms, including the brightly coloured *Tricholomopsis rutilans* (p 17), species of *Paxillus* and *Hypholoma*, and the tough *Lentinus lepideus* (p 20). Other large, fleshy fungi likely to be found, not necessarily mycorrhizal, include species of *Agaricus*, *Clitocybe*, *Collybia*, *Hygrophorus* and *Tricholoma*.

## *Hygrophorus hypothejus*
### Herald of the winter

**Spores**
7–10 ×
4–5 µm,
ellipsoid

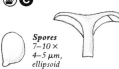

## *H. erubescens*
### Blushing wax agaric

**Spores**
6–10 ×
5–6 µm,
ovoid

× 0.3

A late sp, rarely seen before the frosty weather sets in. Identified by its slimy, yellow-grey cap which has an olive-brown centre and white to orange gills. Prefers acid soil. **Cap** 3–7 cm, convex to depressed, fibrillose. **Gills** decurrent, spaced. **Stem** 4–10 cm, yellowish, slimy, tapers below. *Oct–Nov*.

Variable in size and colour, with a bitter taste. More common in the Scottish Highlands, often in rings. **Cap** 5–10 cm, convex to flat, white with pinkish tints, often yellowish, sticky when wet. **Gills** decurrent, creamy flecked with red, spaced. **Stem** 6–10 cm, yellowish with pink granules near apex. *Sept–Nov*.

## *Clitocybe nebularis*
### Clouded agaric

**Spores**
6–8 ×
3–5 µm,
ellipsoid

Regarded as a good edible sp by many, although it has been known to cause discomfort in some and therefore should initially be treated with caution. Do not confuse with the poisonous *Entoloma sinuatum* (p 133) which has more spaced, pinkish gills. *C. nebularis* grows well in rich soil and has a strong smell of Swedish turnip. **Cap** 6–20 cm, fleshy, convex to depressed, matt or shiny when wet, ash grey to yellow-brown, darker at centre, with white, powdery bloom. **Gills** short-decurrent, cream, crowded. **Stem** 5–12 cm, white to grey. *Aug–Nov*.

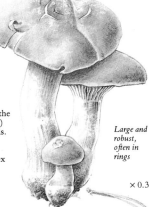

*Large and robust, often in rings*

× 0.3

Coniferous woods

## *Clitocybe clavipes*
### Club-footed clitocybe

***Spores*** 5–7 × 3–4 µm, ellipsoid

× 0.5

The watery flesh of this fungus has a mild taste and scent of bitter almonds. It is recognized by its club-shaped base which is covered in white down, and pale yellow gills. **Cap** 4–8 cm, slightly umbonate to depressed, greyish brown with olive tints, darker at centre, smooth. **Gills** deeply decurrent, thin, soft. **Stem** 5–8 cm, usually paler than cap, with swollen, woolly base. *Sept–Nov.*

## *C. flaccida*
### Tawny funnel cap

***Spores*** 4–5 × 3–4 µm, ovoid, warty

*Earthy smell develops with age*

× 0.5

Often found in clusters or rings near rotting leaf litter, this tawny, thin-fleshed mushroom is edible but not very good. **Cap** 5–10 cm, funnel-shaped with a wavy margin, reddish brown, smooth, shining. **Gills** deeply decurrent, crowded, white to yellow, often spotted brown. **Stem** 3–5 cm, paler than cap, smooth, with woolly base. *Sept–Dec.*

## *C. diatreta*

***Spores*** 3–4 × 2–3 µm

*Thin, flesh-coloured cap*

× 0.5

Very thin and flabby in texture, this species is usually found in groups in pine litter. **Cap** 2–4 cm, convex then flat, often depressed, pale pinkish to tan, darker at centre with thin, wavy margin. **Gills** slightly decurrent, whitish, narrow, crowded. **Stem** 2–5 cm, smooth, colour as cap or paler, with woolly base. *Sept–Nov.*

## *C. langei*

***Spores*** 5–7 × 3–5 µm, broadly ellipsoid

× 0.5

*Strong smell of rancid flour*

One of several closely related spp, *C. langei* can be difficult to identify. It tastes and smells of meal and is especially associated with pines and heathy soils, as found in conifer plantations. **Cap** 2–5 cm, convex to depressed, grey-brown drying whitish from centre, striate when moist. **Gills** decurrent, grey-brown, narrow, crowded. **Stem** 2–4 cm, paler than cap, smooth. *Oct–Dec.*

Coniferous woods

## *Collybia maculata*

### Spotted tough shank

**Spores** 5–6 × 4–5 μm, almost spherical

Pure white when it first appears, this tough-stemmed fungus soon develops rusty spots. The thick white flesh has a mild smell but bitter taste and should not be eaten as it can cause discomfort.
**Cap** 6–12 cm, convex with incurved margin at first, then flat and spotted reddish brown.
**Gills** cream then rust-spotted, adnexed, narrow, crowded.
**Stem** 7–13 cm, firm, grooved. *July–Nov*.

*Tall, fibrous stem with rooting base*

*Large groups embedded in deep litter, frequently under bracken*

× 0.3

## *Tricholoma albobrunneum*

**Spores** 4–5 × 3–4 μm, ellipsoid

Found in rings under pines, *T. albobrunneum* is best avoided, along with other brown *Tricholoma* spp. as it can be very indigestible, even toxic. The mild flesh smells of meal, but has a bitter after-taste. *T. fulvum* (p 97), a birch wood species, is similar but for its pale yellow gills and yellow flesh in the stem. **Cap** 7–10 cm, convex with incurved margin, then flat; chestnut-brown, darker centre, smooth and radially streaked by fine fibrils. **Gills** sinuate, white soon spotted reddish brown, broad, fairly crowded. **Stem** 4–10 cm, dry, cylindrical. *Aug–Nov*.

*Stem is whitish at apex, red-brown below*

*Cap becomes sticky in wet weather*

*Grows well in sandy soil*

× 0.5

Coniferous woods

## *Tricholoma imbricatum*

*Cap surface dry, scaly*

× 0.5

*Slightly rooting base*

**Spores** 5–9 × 4–6 µm, ellipsoid

This mushroom grows well on sandy soil in dense groups. The only other scaly-capped species with which it may be confused is the closely related *T. vaccinum*. The latter is identified by its smaller size, lighter colour and shaggy cap margin. Both species occur in coniferous woodland. **Cap** 4–8 cm, conical to convex, with persistently incurved margin; red-brown surface covered in small, dark, overlapping scales. **Gills** sinuate, fairly crowded, white then pale brown with darker spots. **Stem** 5–9 cm, white then pale brown, fibrillose, tapering above. **Flesh** thick, white, with a mealy smell. *Sept–Nov*.

## *T. portentosum*

**Spores** 5–6 × 3–5 µm, ellipsoid

Appearing in large groups, especially under old pine trees, this sturdy fungus may also be found near frondose trees on acid soil. It has a floury smell and is so good to eat that the French call it "Marvellous Tricholoma" (*Tricholome merveilleux*). **Cap** 4–10 cm, convex then flat, often umbonate; yellowish grey with dark, almost black, fibrils radiating from centre towards paler margin; smooth, sticky when wet. **Gills** sinuate, white to greyish, spaced. **Stem** 6–10 cm, cylindrical, solid, white or discolouring yellowish, with no trace of ring; slightly rooting at base. *Oct–Nov*.

× 0.5

16

Coniferous woods

## *Tricholomopsis rutilans*

### Purple and yellow agaric

*Spores* 7–8 × 5–6 µm, ellipsoid

A brightly coloured agaric with a yellow, bitter-tasting flesh. Grows in clusters near old stumps. **Cap** 5–15 cm, convex or campanulate, expanding to umbonate; yellow surface covered in dense, purple-red scales. **Gills** adnexed, broad, crowded. **Stem** 6–10 cm, pale yellow with purple scales. *Aug–Nov.*

Sulphur-yellow gills

× 0.3

## *T. platyphylla*

### Broad-gilled agaric

*Spores* 7–9 × 5–7 µm, ellipsoid

Found singly on decaying wood, this species is noted for its prominent cord-like mycelium. The thin white flesh is edible but tough and bitter. **Cap** 12–15 cm, convex then flat, smoky brown or paler, translucent when moist, radially streaked with darker fibrils. **Gills** adnexed, very broad, white, spaced. **Stem** 7–13 cm, straight with rooting base. *Sept–Nov.*

Stem white, fibrillose

Mycelial cords well-developed

× 0.25

## *Hygrophoropsis aurantiaca*

### False chanterelle

Dry, velvety surface

*Spores* 5–8 × 3–5 µm, ovoid, slowly dextrinoid

Thin, crowded, forking gills

Often mistaken for Chanterelle (p 103) in colour and shape, the two species are distinguished by their gills, which are not true gills but wrinkled folds in Chanterelle. False chanterelle occurs in large troops and its tough, yellow flesh is best eaten fried, but not recommended. **Cap** 3–6 cm, egg-yellow to orange, convex to deeply depressed with inrolled margin. **Gills** decurrent. **Stem** 2–5 cm, cap colour. *Aug–Nov.*

× 0.5

17

# Coniferous woods

## *Mycena leptocephala*
### Nitrous mycena

## *M. sanguinolenta*
### Small bleeding mycena

## *M. epipterygia*
### Yellow-stemmed mycena

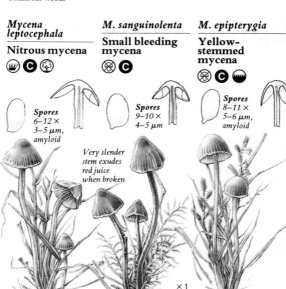

**Spores** 6–12 × 3–5 μm, amyloid

**Spores** 9–10 × 4–5 μm

**Spores** 8–11 × 5–6 μm, amyloid

Very slender stem exudes red juice when broken

Often solitary on soil; smells of nitric acid. **Cap** 0.5–1.5 cm, striate, grey-brown. **Gills** white, ascending. **Stem** 4–6 cm, smooth. *Aug–Nov*.

Amongst leaf litter. **Cap** 1–2 cm, red-brown, campanulate, striate. **Gills** spaced, white, edge tinged. **Stem** 7–10 cm, woolly base. *Aug–Nov*.

**Cap** 1–2 cm, grey to pale yellow-brown, campanulate, striate, slimy. **Gills** white, adnate with decurrent tooth. **Stem** 6–9 cm, slimy. *Aug–Nov*.

## *Marasmius androsaceus*
### Horse-hair fungus

## *Micromphale perforans*

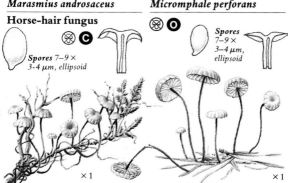

**Spores** 7–9 × 3–4 μm, ellipsoid

**Spores** 7–9 × 3–4 μm, ellipsoid

A fragile sp with a black horse-hair-like mycelium entangling dead needles. **Cap** 0.5–1 cm, slightly depressed, wrinkled, red-brown. **Gills** crowded, white. **Stem** 3–6 cm, black, wiry. *May–Nov*.

Large numbers occur on pine needles; each one grows on a separate needle. Unpleasant smell on bruising. **Cap** 0.5–1.5 cm, pinkish brown, grooved. **Gills** white, narrow, spaced. **Stem** 2–3 cm, black, velvety. *Aug–Oct*.

## Baeospora myosura

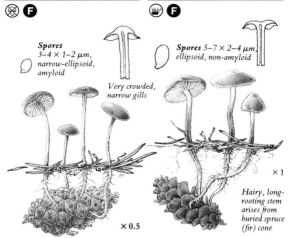

**Spores** 3–4 × 1–2 μm, narrow-ellipsoid, amyloid

Very crowded, narrow gills

× 0.5

Sprouts from buried pine cones. A similar sp, *Strobilurus tenacellus*, has non-amyloid spores and occurs in May. **Cap** 1–2 cm, convex expanding, yellowish to dark brown. **Gills** yellowish, narrow. **Stem** 2–4 cm, slender, with rooting base. *Oct–Dec*.

## Strobilurus esculentus

**Spores** 5–7 × 2–4 μm, ellipsoid, non-amyloid

× 1

Hairy, long-rooting stem arises from buried spruce (fir) cone

Commoner in mountainous spruce forests; only grows on buried fir cones, never on pine. **Cap** 1–2 cm, conical expanding, whitish to yellow-red or brown, shiny. **Gills** adnexed, white, crowded. **Stem** 2–7 cm, pale buff to brown, paler near apex, slender. *Oct–Dec, also May*.

## Xeromphalina campanella

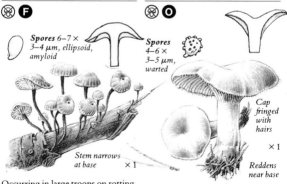

**Spores** 6–7 × 3–4 μm, ellipsoid, amyloid

Stem narrows at base × 1

Occurring in large troops on rotting coniferous wood, *X. campanella* is a fairly rare species, more frequent in mountainous areas. **Cap** 0.5–1.5 cm, convex or campanulate, striate when moist, yellowish rusty brown. **Gills** decurrent, yellow-brown, spaced, with interconnecting veins. **Stem** 1–2 cm, thin, with woolly base. *Aug–Nov*.

## Ripartites tricholoma

**Spores** 4–6 × 3–5 μm, warted

Cap fringed with hairs

× 1

Reddens near base

Found in troops on bare soil, usually under conifers. May be confused with *Clitopilus prunulus* (p 149). **Cap** 2–4 cm, convex to depressed, smooth or radially fibrillose, whitish to pale clay, wavy, with hairy margin. **Gills** white to grey-brown, decurrent, crowded. **Stem** 2–6 cm, whitish, solid. *Sept–Nov*.

Coniferous woods

## *Rhodocybe truncata*

**Spores** 5–8 × 4–6 μm, angular-ovoid

Felty surface

× 0.3

A stocky, fleshy mushroom, usually found growing in clusters or rings. Sometimes mistaken for the toxic *Hebeloma crustuliniforme* (p 87), a less substantial sp with brown spores. **Cap** 4–10 cm, irregular convex, pink-brown to rust, matted surface, with inrolled margin. **Gills** cream to reddish pink, forked, narrow, crowded, sinuate. **Stem** 3–5 cm, stocky, solid, white with pink tints below. *Sept–Nov*.

## *Macrocystidia cucumis*

**Spores** 7–10 × 4–5 μm, ellipsoid

Cap is striate when moist

× 1

Stem tapers below

Often detected amongst sticks and dead leaves by its unpleasant smell of fish or cucumber. **Cap** 1–4 cm, campanulate, dark brown drying paler near margin. **Gills** pale yellow-pink, thin, crowded. **Stem** 4–6 cm, dark brown, paler near apex, velvety. *Sept–Nov*.

## *Lentinus lepideus*

**Spores** 9–14 × 3–5 μm, cylindric

× 0.3

Gills have a toothed edge

Always on coniferous wood, this sp will even grow on creosoted railway sleepers. **Cap** 5–10 cm, pale yellow, convex to depressed, firm, cracked in concentric zones of brown scales. **Gills** whitish, broad. **Stem** 2–8 cm, hard, white, scales brown. *May–Oct*.

## *L. tigrinus*

**Spores** 6–8 × 2–3 μm, cylindric

× 0.5

Smaller, finer scales than in L. lepideus

Slender stem narrows towards base

A smaller, more slender fungus than *L. lepideus*; also grows on deciduous wood. **Cap** 2–10 cm, whitish with concentric zones of tiny, fibrillose, dark scales. **Gills** cream, decurrent. **Stem** 2–5 cm, scaly. *May–Oct*.

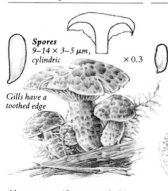

## Phyllotopsis nidulans

*Spores 5–6 × 2–3 μm, curved-cylindric*

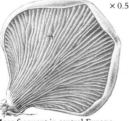

×0.5

More frequent in central Europe than in Britain. Laterally attached to stumps (also beech), its bright orange gills distinguish it from *Paxillus panuoides* (p 39), which has brown gills and spores. **Cap** 2–5 cm, convex, almost shell-shaped, yellowish to pale orange, velvety, with incurved margin. **Gills** bright orange-yellow, crowded, narrow. **Stem** absent. **Flesh** yellow, firm, spongy. *Oct–Dec.*

## Panellus mitis

*Spores 3–5 × 1–2 μm, amyloid*

×0.5

*Large numbers of this white fungus may be seen on felled wood*

Like small white mussels attached to dead wood, especially spruce twigs, this species may be quickly identified by a jelly-like strip on the edge of the gills which can be peeled off. **Cap** 1–2 cm, shell- or kidney-shaped with a short lateral stem, white discolouring to clay brown, with a separable pellicle. **Gills** white, fairly crowded. **Stem** 0.3–0.5 cm, white, powdery, flattened and dilated where it joins the cap. **Flesh** spongy with a mild taste, but not worth eating. *Oct–Feb.*

## Lyophyllum fumatofoetens

*Spores 6–9 × 2–4 μm, narrow-ellipsoid, rough*

Also widely known as *Collybia gangraenosa* and *Tricholoma leucophaeatum*. The flesh blackens on exposure, and the whole fruitbody will turn black if there is a lack of moisture. Differs from other related spp which blacken by its rough spores. **Cap** 4–9 cm, convex, sooty grey, fibrillose, with a shaggy margin. **Gills** pale yellowish grey staining brown-black, adnexed, crowded. **Stem** 5–9 cm, cylindric, streaky, cap colour. *Sept–Nov.*

*All parts bruise black very easily*

×0.5

Coniferous woods

## *Cystoderma amianthinum*
### Saffron parasol

## *C. carcharias*
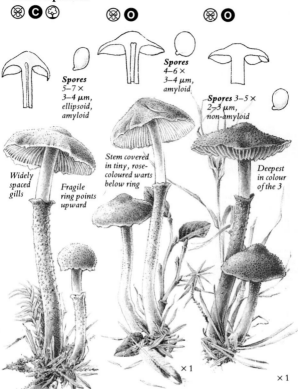

## *C. granulosum*

*Spores 5–7 × 3–4 μm, ellipsoid, amyloid*

*Spores 4–6 × 3–4 μm, amyloid*

*Spores 3–5 × 2–3 μm, non-amyloid*

*Widely spaced gills*

*Fragile ring points upward*

*Stem covered in tiny, rose-coloured warts below ring*

*Deepest in colour of the 3*

× 1

The Saffron parasol is the commonest of the three *Cystoderma* spp shown here and it may also be found in mossy heaths and mixed woods. The thin yellow flesh has an unpleasant, mouldy smell. A chemical test is to treat the pellicle with 10% caustic potash, which should turn rusty brown. **Cap** 2–5 cm, convex, often umbonate, with dry, ochre-yellow, granular surface and shaggy margin. **Gills** adnate, white. **Stem** 4–8 cm, granular up to ring, then smooth. *Aug–Oct.*

Occurs amongst moss and short grass, often under pine and spruce. The thin white ring is usually more prominent than that of the Saffron parasol and the pellicle does not turn brown with caustic potash. **Cap** 2–4 cm, campanulate expanding, often umbonate, pinkish grey, dry, granular, with shaggy margin. **Gills** white, adnate, narrow. **Stem** 4–5 cm, cap colour, finally brownish, with granular surface below ring, created by the remnants of the outer veil. *Aug–Nov.*

This sp is recognized by its red-brown colour and the hoary sheen of the granular veil. The pellicle turns rusty brown when treated with caustic potash. **Cap** 3–6 cm, convex to umbonate, often wrinkled; reddish brown turning rusty yellow with age, dry and granular. **Gills** slightly adnexed, creamy yellow, broad. **Stem** 4–6 cm, colour as cap, thickened slightly towards base; coarsely granular up to membranous ring, then smooth and pale. **Flesh** thin, whitish, rarely tinged red. *Aug–Oct.*

Coniferous woods

## *Lactarius deliciosus*

### Saffron milk cap

Thick, fleshy and edible, although not as delicious as it sounds. Its sweet, orange milk has an acrid after-taste. **Cap** 6–20 cm, orange-red with darker, concentric zones, bruising green; convex to depressed, sticky, with thin, inrolled margin. **Gills** short-decurrent, narrow, crowded. **Stem** 2–7 cm, short, thick, with orange spots. **Flesh** off-white, discolours slowly. *Aug–Oct.*

*Bright orange gills become spotted green with age*

× 0.3

## *L. deterrimus*

Not as good to eat as *L. deliciosus*, it is distinguished by its unmarked stem. More common in Scandinavia, near spruce. **Cap** 10–15 cm, convex to depressed with incurved margin; reddish orange with faint, crowded, narrow zones near edge; sticky, becoming green with age. **Gills** crowded, decurrent. **Stem** 3–10 cm, stout, lacking spots. *Aug–Sept.*

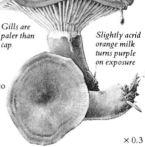

*Gills are paler than cap*

*Slightly acrid orange milk turns purple on exposure*

× 0.3

## *L. scrobiculatus*

The thick, hard flesh produces an acrid white milk which turns yellow. Common in damp woods of Scandinavia. **Cap** 10–20 cm, convex then centrally depressed, straw yellow with faint zones, slimy when wet. **Gills** light yellow, decurrent, thin, crowded. **Stem** 10–15 cm, cylindric, with conspicuous yellow spots. *Sept–Nov.*

*Shaggy margin*

× 0.3

## *Lactarius camphoratus*
### Curry-scented milk cap
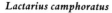  

## *L. rufus*
### Rufous milk cap
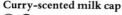

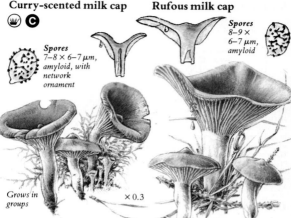

**Spores** 7–8 × 6–7 μm, amyloid, with network ornament

**Spores** 8–9 × 6–7 μm, amyloid

Grows in groups  × 0.3

Prefers non-chalky soil  × 0.3

A small, slender, red-brown sp with a white, watery, mild-tasting milk. When dried it has a strong, spicy smell, and may be used as a seasoning. The gills and stem tinge purple with age. **Cap** 2–6 cm, convex to depressed with a small umbo, red-brown, with a furrowed margin. **Gills** decurrent, paler than cap. **Stem** 3–5 cm, cap colour. *Aug–Nov.*

In eastern Europe it is preserved in salt and eaten fried. The white milk does not change colour and has a hot, acrid taste. **Cap** 4–10 cm, dark red-brown, smooth, convex to depressed, with a pointed umbo. **Gills** short-decurrent, pale yellow then reddish, crowded. **Stem** 5–8 cm, whitish at base. *Aug–Nov.*

## *L. helvus*

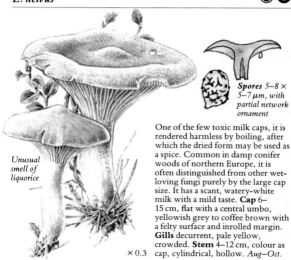

**Spores** 5–8 × 5–7 μm, with partial network ornament

Unusual smell of liquorice  × 0.3

One of the few toxic milk caps, it is rendered harmless by boiling, after which the dried form may be used as a spice. Common in damp conifer woods of northern Europe, it is often distinguished from other wet-loving fungi purely by the large cap size. It has a scant, watery-white milk with a mild taste. **Cap** 6–15 cm, flat with a central umbo, yellowish grey to coffee brown with a felty surface and inrolled margin. **Gills** decurrent, pale yellow, crowded. **Stem** 4–12 cm, colour as cap, cylindrical, hollow. *Aug–Oct.*

Coniferous woods

## *Russula adusta*

*Greyish flesh does not redden when cut*

*Cap surface is sticky and shiny*

**Spores** $7–9 \times 6–8$ μm, *amyloid, with small warts*

The hard flesh smells of old wine and is edible but not recommended. May be mistaken for *R. densifolia* which bruises red. As with other *Russula* spp, *R. adusta* does not release a milk when damaged. **Cap** 5–20 cm, convex expanding, greyish white discolouring dark grey-brown. **Gills** decurrent, creamy, crowded. **Stem** 3–8 cm, stout, white to grey. *Aug–Oct.*

× 0.25

## *R. consobrina*

**Spores** $9–11 \times 8–9$ μm, *warty with connectives*

*Thick white gills*

Similar to *R. sororia*, an oakland sp with a grooved cap margin. *R. consobrina* has a smooth cap and is often found under spruce. It has an acrid taste. **Cap** 7–10 cm, convex to depressed, grey-brown, sticky. **Gills** adnexed, crowded. **Stem** 5–8 cm, grey-white. *Aug–Oct.*

× 0.3

## *R. emetica*

### Emetic russula

**Spores** $9–11 \times 8–9$ μm

*Cap is scarlet in peak condition*

*Gills and stem both white*

Has a hot, acrid taste and causes vomiting if eaten raw. Found in damp, mossy places, this sp may be confused with *R. paludosa* (p 26) and *R. lepida* (p 112), which have cream (not white) gills. The slimy pellicle of *R. emetica* peels easily, revealing thin red flesh. **Cap** 5–9 cm, light red to cherry red, fading to pink when wet, convex to depressed. **Gills** adnate, quite spaced. **Stem** 3–8 cm, thicker at base. *July–Oct.*

× 0.3

Coniferous woods

## *Russula paludosa*

*Cap is shiny apple red when dry*

× 0.25

**Spores** 8–10 × 7–8 μm, with large warts, amyloid

This red russula, the largest of its kind, is most likely to be found in peaty areas of Scotland. In Finland it is sold commercially. **Cap** 6–20 cm, convex to depressed, orange to blood red, sometimes yellow at centre, sticky. **Gills** pale cream, finally yellow. **Stem** 5–15 cm, white, often flushed pink. *July–Sept.*

## *R. queletii*

*Fruity but unpleasant smell*

× 0.3

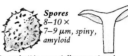

**Spores** 8–10 × 7–9 μm, spiny, amyloid

*R. queletii* is especially associated with spruce in damp places, but its acrid taste makes it unsuitable for the table. The white, fragile flesh is pink just under the pellicle. **Cap** 3–8 cm, dark purplish red to violet, almost black at centre, flushing dirty olive with age; convex soon depressed, with a grooved margin. **Gills** white, becoming lemon yellow, fairly crowded. **Stem** 4–6 cm, cylindric, pink to pale violet. *July–Oct.*

## *R. azurea*

× 0.5

**Spores** 8–10 × 7–8 μm, with low warts

A comparatively rare sp, *R. azurea* is restricted to spruce forests, especially in mountainous areas. It resembles *R. parazurea* (p 113) which grows under beech and oak, but its gills are pure white rather than cream-coloured. The white flesh lacks any distinctive smell but has a mild flavour none the less, and is worth gathering if sufficient quantities can be found. **Cap** 4–8 cm, dark blue-violet to lilac grey, convex expanding to flat, finally depressed, with a loose, cottony surface. **Gills** pure white, spaced. **Stem** 3–5 cm, white. *July–Oct.*

## *R. caerulea*

*Cap always has a central umbo*

**Spores** 8–10 × 7–9 μm, warty

The flesh of this sp actually has a mild taste, but the whole fungus is considered inedible because of its very bitter pellicle. **Cap** 5–7 cm, almost conical at first, expanding to depressed, typically with an umbo; dark violet to purplish red, blackish at centre, shiny, slippery when wet. **Gills** lemon to ochre, fairly crowded. **Stem** 4–8 cm, white to grey, with a swollen base. *June–Oct.*

×0.5

## *Entoloma nitidum*  *Nolanea cetrata*  *N. cuneata*

**Spores** 7–11 × 6–9 μm, ovoid-angular

**Spores** 9–12 × 6–8 μm

**Spores** 9–13 × 7–9 μm

*Cap often slightly scaly in the centre*

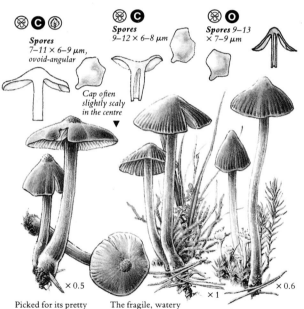

×0.5   ×1   ×0.6

Picked for its pretty deep blue colour, but probably best avoided. Found on peaty soil. **Cap** 2–5 cm, conical expanding with a pointed umbo, dark indigo blue, silky. **Gills** white then pink, ascending. **Stem** 6–9 cm, fibrillose-striate, with white, rooting base. *Sept–Nov.*

The fragile, watery flesh lacks a distinct smell. Common in damp, mossy places. **Cap** 1–3 cm, convex to depressed, yellow-brown, translucent and striate when moist, but opaque when dry. **Gills** yellowish pink, spaced. **Stem** 4–7 cm, paler than cap, silky-striate. *Sept–Nov.*

More frequent in Scotland. **Cap** 2–5 cm, sepia brown with persistent yellow umbo, drying pale brown and silky-shiny; conical, scarcely expanding. **Gills** free, yellow-brown then pink. **Stem** 8–11 cm, paler than cap, fibrous. *Oct–Nov.*

Coniferous woods

## *Galerina unicolor*

*Stem darker below ring*

**Spores** 8–10 × 5–7 µm, with a loose perisporium

× 0.6

Grows in tufts on conifer stumps. **Cap** 1.5–3 cm, dark brown drying ochre yellow, with darker, striate margin; convex to flat, smooth, sticky. **Gills** cinnamon brown, narrow, adnate. **Stem** 2–6 cm, slender, pale above membranous ring, dark below. *Sept–Nov*.

## *Hebeloma mesophaeum*

**Spores** 8–10 × 5–6 µm, ellipsoid, faintly warty

The flesh is dark brown in the stem base. **Cap** 2–4 cm, convex to flat, slimy at first, shiny; clay to chestnut brown, paler towards margin. **Gills** sinuate, pinkish brown to cinnamon. **Stem** 4–7 cm, slender, fibrillose, white soon tinted brown, with distinct ring zone.
× 0.5  *Aug–Nov*.

## *Phaeocollybia festiva*

**Spores** 7–9 × 4–5 µm, ovoid, faintly warty

*Whole fruitbody slimy to touch*

*Long rooting base*

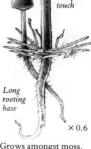

× 0.6

Grows amongst moss. **Cap** 2.5–4 cm, conical or campanulate with a pointed umbo, olive brown drying greenish from the centre. **Gills** adnate, pink-brown. **Stem** 6–10 cm, dark brown, smooth, hollow. *Sept–Nov*.

## *Gymnopilus penetrans*

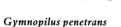 

*Smooth, dry cap surface*

**Spores** 6–9 × 4–5 µm, warty

Large numbers occur on dead wood. The yellow flesh has a bitter taste. **Cap** 3–6 cm, radially fibrillose, convex to umbonate, chrome yellow to rust brown. **Gills** yellow, staining rust brown, thin, fairly crowded. **Stem** 3–6 cm, cap colour, but paler near apex and whitish at base, fibrous, with remnants of yellow veil in ring zone. *Aug–Nov*.

× 0.5

*Grows on fallen twigs and stumps*

Coniferous woods

## *Pholiota flammans*

**Spores** $3-4 \times 2-3\ \mu m$

Very scaly

This striking yellow fungus is common in Scotland, where it grows in tufts on decaying stumps. **Cap** 2–6 cm, convex expanding, dry, golden yellow with recurved, sulphur-yellow scales. **Gills** adnexed, yellow, crowded. **Stem** 3–6 cm, colour as cap, often curved, covered with recurved scales up to scaly, yellow ring, then smooth. **Flesh** firm, yellow. *Sept–Nov.*

× 0.5

## *Cortinarius collinitus*

**Spores** $12-20 \times 7-9\ \mu m$, almond-shaped, warty

Glutinous cap

Gills are amethyst at first

The stem of this tall, slimy species is attractively marked by violet, scaly bands left by the veil. Usually in groups in deep moss. **Cap** 3–10 cm, convex with low umbo, tawny orange, darker centre. **Gills** adnate, pale clay brown or grey-blue, finally rust brown, spaced. **Stem** 5–12 cm, yellow-brown with a violet slime which forms scaly bands. *Aug–Oct.*

× 0.3

## *C. mucosus*

**Spores** $13-16 \times 6-7\ \mu m$, almond-shaped, warty

Both *C. collinitus* (above) and *C. mucosus* belong in the subgenus *Myxacium*, a group of at least 20 spp in Europe which all have a slimy veil on the cap and stem when young. *C. mucosus* is a fairly rare fungus which has a preference for sandy and heathy soils, especially under pines. **Cap** 4–10 cm, convex expanding, chestnut brown, very slimy when damp. **Gills** white to cinnamon brown, adnate, quite spaced. **Stem** 5–15 cm, almost white, cylindric, sticky, not scaly. *Sept–Nov.*

Slimy, chestnut-brown cap

× 0.3

Coniferous woods

## *Cortinarius purpurascens*

**Spores**
$8–10 \times 5–6 \mu m$, ellipsoid, warty

Species of the genus *Cortinarius* which have a slimy cap and dry stem, even in wet weather, are grouped into the subgenus *Phlegmacium*. *C. purpurascens* is such a species and is almost always found tufted on non-chalky soil. **Cap** 5–15 cm, convex to broadly umbonate, date brown varying to olive brown, often with darker streaks and a deep violet margin which becomes paler; slimy when young. **Gills** violet then rusty brown, bruising deep purple. **Stem** 5–12 cm, stout, fibrous, with a bulbous base which often has a distinct rim; violet, with remains of purplish cortina near apex. **Flesh** violet. *Sept–Nov*.

× 0.3

## *C. speciosissimus*

**Spores**
$9–12 \times 6–9 \mu m$, broadly ellipsoid

For years, *Cortinarius* spp were thought to be a harmless group of fungi, but there are a few amongst them, *C. speciosissimus* included, which are now known to have caused many untimely deaths. The poisonous effects of this sp are usually not felt until after 3–14 days, by which time it is often too late to remove the toxins from the body. Fortunately, this sp is rare in Britain. **Cap** 2–8 cm, convex expanding with a pointed umbo, tawny red with yellowish margin, covered with minute, appressed scales. **Gills** bright ochre to tawny rust, broad, widely spaced. **Stem** 5–11 cm, thick, cylindrical, pale ochre then rusty, with yellow patches. **Flesh** ochre yellow, strong smell of radish. *Sept–Nov*.

*Stem is fibrillose striate, often with yellow bands left by the veil*

× 0.5

Coniferous woods

## C. callisteus

***Spores** 6–8 × 5–7 µm, broadly ovoid, warty*

A large, golden-yellow fungus, often amongst moss. **Cap** 3–9 cm, convex to slightly depressed, bright tawny yellow to rust brown, dry, with tiny, fibrous scales. **Gills** deep yellow then tawny brown, fairly well spaced. **Stem** 5–15 cm, variable in shape, from cylindric to very swollen and bulbous at base; colour as cap, with pale yellow, scaly remnants of veil near middle; striate. *Aug–Oct.*

*Stem base is very swollen*

× 0.5

## C. traganus

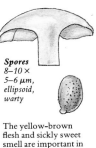

***Spores** 8–10 × 5–6 µm, ellipsoid, warty*

The yellow-brown flesh and sickly sweet smell are important in separating *C. traganus* from other bluish *Cortinarius* spp. More frequent in Scotland. **Cap** 4–12 cm, convex to broadly umbonate, fleshy, dry, silky-scaly; pale blue-violet, then yellowish to rusty ochre. **Gills** pale ochre to rust, adnexed, fairly crowded. **Stem** 6–12 cm, stout, lilac then pale ochre from base up; remnants of cortina visible in upper part. *July–Oct.*

*Surface may develop cracks*

*Gills are discoloured rusty brown by the spores*

*Young fruitbody amethyst blue, with hemispherical cap joined to stem by cobweb-like veil*

*Bulbous, club-shaped stem base*

× 0.5

31

Coniferous woods

## *Cortinarius decipiens*

**Spores**
8–9 ×
5–6 μm,
ellipsoid,
warty

Found in damp places and occasionally under elder and birch. *C. decipiens* falls within the subgenus *Hydrocybe*, a group typified by change in cap colour with moisture and lack of a distinct ring zone on the stem. **Cap** 1–4 cm, campanulate expanding, always with an umbo, bay brown drying pale tan. **Gills** cinnamon, thin, crowded. **Stem** 5–7 cm, pale, tinted lilac. *Sept–Nov.*

*Pale, shiny stem*

× 0.6

## *C. glandicolor*

**Spores**
8–9 ×
4–5 μm,
ellipsoid,
warty

*Surface dry and silky*

Difficult to identify as there are many similar spp. Grows in groups in boggy areas of pine and spruce woods. **Cap** 2–7 cm, convex with conical umbo, thin, dark red-brown drying paler, often splitting. **Gills** adnate, brown, spaced. **Stem** 4–6 cm, slender, striate, dark brown with white ring zone. *Aug–Nov.*

× 0.6

## *C. gentilis*

*Usually occurs in troops*

**Spores**
8–9 ×
5–6 μm,
ovoid,
warty

*C. gentilis* and *C. glandicolor* (above) both belong to subgenus *Telamonia*, a large group of mostly small-sized *Cortinarius* spp, with a cap that dries paler and distinct ring zones on the stem. A similar sp, *C. saniosus*, of frondose woods (often with poplar), has a tawny stem with a simple, cobweb-like ring zone. The yellow zones on the stem of *C. gentilis* are distinctive. **Cap** 1–3 cm, conical to campanulate, expanding, silky, tawny brown drying golden yellow. **Gills** yellow to brown, thick, spaced. **Stem** 3–6 cm, yellow-brown, curved, silky. *Aug–Nov.*

× 1

Coniferous woods

## *Dermocybe cinnamomea*

### Cinnamon cortinarius

**Spores**
6–8 ×
3–4 µm,
ellipsoid, warty

*Cap colour varies from bright red-brown to yellowish olive brown*

This sp has bright lemon-yellow gills when young and a dry, silky cap that does not change colour when wet. May occur in large numbers, often tufted. The thin flesh is lemon to chrome yellow. Similar spp include *D. cinnamomeo-lutescens* and *D. cinnamomeo-badius*. The former has persistent lemon-yellow gills and a yellower cap, while the latter has a chestnut-brown cap with a yellow edge. **Cap** 2–6 cm, convex to umbonate with incurved edge. **Gills** orange or golden red, crowded. **Stem** 5–8 cm, with brown fibrils. *Sept–Nov.*

× 0.5

## *D. sanguinea*

### Blood-red cortinarius

**Spores**
7–8 ×
4–5 µm,
ellipsoid,
warty

*Surface has a fine, scaly texture*

*Cap, gills and stem are blood red*

Even the flesh of this rather vivid sp is red. It is only likely to be confused with *D. cinnabarina* which is wholly scarlet and found in beech woods. **Cap** 2–5 cm, convex to umbonate at first, then flat or depressed. **Gills** adnate, crowded. **Stem** 6–10 cm, slender, cap colour or darker, fibrillose. *Sept–Nov.*

× 0.5

## *D. semisanguinea*

**Spores**
6–8 ×
4–5 µm,
ellipsoid,
warty

*Dry, rusty-yellow cap*

*Dermocybe* spp are recognized by a non-sticky cap which does not dry paler, a slender stem and thin flesh. *D. semisanguinea* differs from *D. cinnamomea* by its blood-red gills. **Cap** 2–5 cm, convex expanding to umbonate, rusty yellow to rich tawny brown, silky. **Gills** blood red, crowded. **Stem** 3–5 cm, yellowish brown, often with rusty fibrils. **Flesh** yellowish brown. *Sept–Nov.*

× 0.5

Coniferous woods

## *Rozites caperata*

*Cap covered by loose, cottony veil*

*Ring on stem is thick, fleshy and persistent*

**Spores** 13–15 × 7–9 μm, almond-shaped, warty

More frequent in the mountainous areas of Scandinavia and central Europe, especially on acid soils. Has a mild taste and is much sought after, but soon attacked by insects. **Cap** 4–12 cm, strongly convex expanding, dry, thick, dull ochre yellow to rusty brown, orange when moist. **Gills** adnate, light brown to rust-coloured, thick, crowded. **Stem** 8–12 cm, cylindric, firm, cream, striate; ring fleshy. *Aug–Oct.*

× 0.3

## *Psathyrella caput-medusae*

*Cap has a dark brown centre*

*Lower half of stem covered with brown, recurved scales*

*Ring has a double layer*

*Grows in tufts on and around conifer stumps*

**Spores** 10–12 × 4–6 μm, with germ-pore

Usually recognized by its scaly white cap and thick, flaring ring. These features are not typical of the genus *Psathyrella*, a large group of about 70 spp in Britain alone, most of which are small and delicate. The flesh has a strong, sweet smell. **Cap** 4–7 cm, convex expanding, white with dark brown centre and zones of large, brown, fibrillose scales. **Gills** adnexed, grey then dark grey-brown, crowded. **Stem** 5–7 cm, thick, whitish with recurved, brown scales below conspicuous, fibrillose ring; smooth above. *Sept–Oct.*

× 0.6

Coniferous woods

## *Agaricus silvicola*
### Wood mushroom

*Cap is never sticky*

*Spores 5–6 × 3–4 µm, ellipsoid*

May be confused with the deadly *Amanita* species when young, so stem and gills should be examined with care. It smells of aniseed.
**Cap** 6–12 cm, strongly convex, expanding, with a thin margin, creamy white bruising yellow, dry, smooth. **Gills** pale grey then flesh brown, finally chocolate, very crowded. **Stem** 6–12 cm, silky-smooth. *July–Nov.*

× 0.5

*Ring is persistent but fragile, hangs down*

*Stem has a bulbous base, but no volva*

*Grows on ground, usually in clearings*

## *A. purpurellus*

## *A. comtulus*

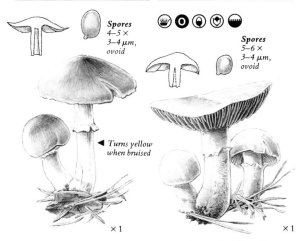

*Spores 4–5 × 3–4 µm, ovoid*

*Spores 5–6 × 3–4 µm, ovoid*

◄ *Turns yellow when bruised*

× 1

× 1

Occurs either singly or in small groups amongst pine needles. Differs from the more typical *Agaricus* spp by its small size, thin flesh and slender stem. Smells of bitter almond. **Cap** 2–4 cm, convex soon flat, purplish or reddish with a darker centre, silky, radially fibrillose. **Gills** free, grey then pink, crowded. **Stem** 3–4 cm, white; ring thin, near middle. *July–Sept.*

Looks like a small Field mushroom, *A. campestris* (p 152), but the bright pink gills are distinctive. Also occurs in meadows, especially near the sea. **Cap** 2–4 cm, fleshy, convex or flat, white slightly yellowish when old, silky. **Gills** free, at first bright rose or flesh pink, finally dark brown, crowded. **Stem** 2–3 cm, thicker at base, white or pale yellow, with thin, narrow ring. *Sept–Nov.*

## *Agaricus silvaticus*

*Spores*
*4–6 ×*
*3–4 μm,*
*ovoid*

The thin, white flesh reddens when cut but in older specimens the flesh is brownish and does not discolour. May be distinguished from *A. langei* (below) by its bulbous base.
**Cap** 6–10 cm, strongly convex, expanding, covered with dense, red-brown, appressed, fibrillose scales on a white background.
**Gills** pinkish grey becoming blackish brown, crowded.
**Stem** 7–11 cm, whitish, rather scaly, soon hollow, with a swollen base and large, thin, persistent ring. Sept–Nov.

*Deeply rooted in pine needles*

× 0.5

## *A. langei*

*Flesh discolours blood red immediately on exposure*

*Scales are more pronounced than in A. silvaticus*

*Spores*
*7–8 × 4–5 μm,*
*oblong-ovoid*

A variable sp in size and colour, *A. langei* grows in groups under spruce, sometimes under oak. The cap has distinct rings of flat, dark brown scales and the white flesh stains far more intensely than that of *A. silvaticus*.
**Cap** 6–12 cm, convex expanding, rust brown, darker at centre, covered in fibrillose scales. **Gills** rosy pink then blackish brown, crowded, free.
**Stem** 7–12 cm, white, slightly scaly, cylindric, with thick white ring; bruises red very easily. *Aug–Nov.*

*Large, white ring has dark brown scales on the underside*

*Stem not bulbous*

× 0.5

Coniferous woods

## A. abruptibulbus

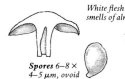

*White flesh smells of almond*

*Silky white cap bruises yellow*

**Spores** 6–8 × 4–5 μm, ovoid

Small groups may be found embedded in pine needles. It is very similar to *A. silvicola* (p 35), but for its more flattened, bulb-like stem base which has a distinct rim. **Cap** 8–12 cm, campanulate expanding to broadly convex, finally flat, silky, slightly scaly at the margin. **Gills** greyish pink to blackish brown, crowded, free. **Stem** 10–12 cm, white, smooth, with a broad, white ring which hangs down. *July–Nov*.

*Base of stem is abruptly swollen*

× 0.5

## Amanita porphyria

**Spores** 8–9 × 7–8 μm, amyloid

*Stem soon becomes hollow*

Not poisonous but still unpleasant, this sp is one to be avoided for risk of confusion with the Death cap (p 104). It is usually distinguished from other *Amanita* spp by the grey-brown colour of the cap, which usually has a purple tinge. It occurs only in coniferous woods, never in fields or open spaces, and is more common in northerly areas. **Cap** 3–8 cm, campanulate expanding to convex, then flat, often with a low umbo; greyish brown to rusty brown, darker at the centre; surface smooth and shiny, bearing a few, small, white scales, the remnants of the veil. **Gills** free, white, crowded. **Stem** 4–10 cm, tall, thin, white to violet-grey, with a swollen base cupped by a loose, membranous volva; ring high, hanging down, white then brown, greyish on the underside, striate. *July–Oct*.

*Cap margin smooth, not striate*

*Ring darkens with age*

× 0.6

Coniferous woods

## *Hypholoma capnoides*

*Gills never yellow*

*Spores 7–9 × 4–5 µm, with germ-pore*

In tufts on stumps. **Cap** 1–8 cm, convex or flat, ochre yellow, paling towards edge, smooth, sticky when damp. **Gills** bluish grey to dark brown, fairly crowded. **Stem** 3–9 cm, slender, hollow, pale yellow tinged brown from base up. *Sept–Nov.*

*Silky, shiny stem*  × 0.3

## *H. marginatum*

*Spores 7–9 × 4–5 µm, ovoid with germ-pore*

Also known as *H. dispersum*. Occurs on the ground and amongst old roots. The cap edge is white, often with remnants of a black, cobweb-like veil. Too bitter to eat. **Cap** 1–4 cm, convex or campanulate, olive yellow, darker at centre, smooth, shiny, dry. **Gills** adnate, yellow to purplish black. **Stem** 3–6 cm, tough, rusty at base. *July–Oct.*

*Thin stem whitish at apex*

*Scales give delicate snake-skin effect*

× 0.5

## *Stropharia hornemannii*

*Spores 10–13 × 5–7 µm, ellipsoid, with germ-pore*

The genus *Stropharia* comprises mostly medium- to small-sized spp with a distinct membranous ring, adnate gills and purple-brown spores. Of these, *S. hornemannii* is the largest, and its fleshy form may be seen growing in or near swamps and peat bogs. It has an unpleasant taste and smell. **Cap** 4–10 cm, strongly convex, sticky when young, smooth, pale yellow-brown to chestnut. **Gills** greyish to violet-brown, crowded. **Stem** 8–12 cm, white to yellowish, granular below thick, fibrillose ring. *Sept–Nov.*

× 0.3

Coniferous woods

## *Paxillus atrotomentosus*

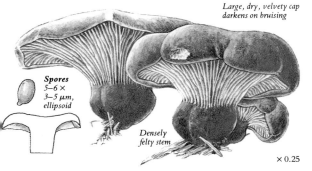

*Large, dry, velvety cap darkens on bruising*

**Spores** 5–6 × 3–5 µm, ellipsoid

*Densely felty stem*

× 0.25

The short, stout stem with its brown-black felty surface soon identifies this species. The flesh is thick, soft and creamy but its persistent bitter taste makes it inedible. Grows on or near conifer stumps, especially pine and spruce. **Cap** 5–30 cm, yellowish chestnut brown, convex soon flat, often depressed at centre, with inrolled margin. **Gills** decurrent, ochre yellow spotting brown, thin, forked, soft and very crowded; easily separated from cap flesh. **Stem** 4–9 cm, central or excentric, robust, solid. *July–Oct.*

## *P. panuoides*

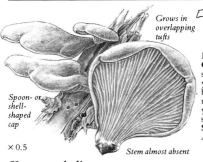

*Grows in overlapping tufts*

**Spores** 5–6 × 3–5 µm, ellipsoid

*Spoon- or shell-shaped cap*

× 0.5

*Stem almost absent*

Edible but mediocre. **Cap** 3–12 cm, variable shape, dry, greenish yellow spotted rust brown, with inrolled margin. **Gills** orange-yellow, narrow, forking, separable, crowded. **Stem** laterally attached. *Aug–Nov.*

## *Xerocomus badius*

### Bay boletus

**Spores** 12–15 × 4–5 µm, spindle-shaped

This sp has a large chestnut-brown cap and lemon-yellow pores which turn blue-green when touched. The white flesh discolours bluish. On acid soils. **Cap** 5–20 cm, hemispherical expanding to almost flat; velvety and shiny when dry, sticky when wet. **Tubes** adnate, pale yellow turning greenish. **Pores** angular. **Stem** 5–12 cm, robust, paler than cap, fibrillose. *Aug–Nov.* × 0.25

*No network on stem*

## Suillus bovinus

**Spores** 6–11 × 2–4 μm

Slimy or sticky-shiny

Large, greenish-yellow pores

× 0.25

Although edible, the thick spongy flesh is gelatinous and not recommended. The angular pores, if examined closely, will be seen to be subdivided by sunken walls into 3 or 4 smaller pores. Found in large numbers in damp situations under pines. **Cap** 4–15 cm, convex expanding, often umbonate, yellowish brown to tawny. **Tubes** slightly decurrent, olive brown. **Pores** radially arranged, not bluing. **Stem** 4–8 cm, curving, often tapering below, solid, colour as cap or paler. **Flesh** creamy white to yellow. *July–Nov.*

## S. granulatus    S. variegatus

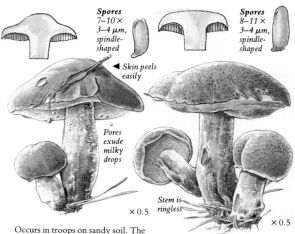

**Spores** 7–10 × 3–4 μm, spindle-shaped

◀ Skin peels easily

Pores exude milky drops

× 0.5

**Spores** 8–11 × 3–4 μm, spindle-shaped

Stem is ringless

× 0.5

Occurs in troops on sandy soil. The stem has a distinctive granular surface near the apex which darkens with age. Best eaten deep fried, with the pellicle removed. **Cap** 5–15 cm, hemispherical then convex, smooth, glutinous, pale yellow to red-brown. **Tubes** adnate, fine, yellow. **Pores** irregular, small, dense, pale yellow. **Stem** 5–10 cm, solid, fairly soft, whitish then pale yellow, darker at base. **Flesh** white to sulphur yellow. *Aug–Oct.*

The cap is sticky only in wet weather; otherwise it is felty and cracked into tiny rust-coloured scales. The spongy yellow flesh has an unpleasant smell of earthballs (*Scleroderma*) and may turn blue when cut. **Cap** 5–12 cm, convex expanding to depressed, yellow-brown. **Tubes** adnate, olive yellow. **Pores** angular, small, bluing slowly on bruising. **Stem** 4–11 cm, firm, colour as cap, striate. *June–Nov.*

Coniferous woods

## *S. aeruginascens*

***Spores**
9–12 ×
4–5 µm,
spindle-
shaped*

Slimy surface

Sticky ring erodes with age, leaving brown zone

Also known as *Boletus viscidus*. This species is always associated with larch. Edible but mediocre; peel off the slimy skin before eating. **Cap** 4–7 cm, convex expanding, shiny, beige to reddish grey, often slightly ridged. **Tubes** adnate to decurrent, greyish. **Pores** large, irregular. **Stem** 5–8 cm, cap colour or paler with brown streaks; ring high, white or brown, points up. **Flesh** white, tinted pink. *June–Oct.*

× 0.5

## *S. grevillei*
## Larch boletus

## *S. luteus*

***Spores**
7–10 ×
2–4 µm*

***Spores**
7–10 ×
3–4 µm,
spindle-
shaped*

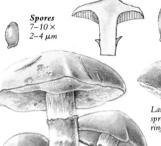

No granules on stem

× 0.5

Large, spreading ring

Stem covered in brownish granules

× 0.5

Also known as *Boletus elegans*. Grows in grass exclusively under larch. Considered edible but can be slightly indigestible. **Cap** 5–10 cm, convex, golden yellow to rusty orange, slimy, smooth. **Tubes** adnate to decurrent, short, yellow. **Pores** round then angular, greyish yellow to bright yellow, staining red-brown. **Stem** 5–10 cm, solid, fibrillose, darker than cap with slimy, upward-pointing ring. **Flesh** bright yellow. *July–Oct.*

Grows well on sandy soil, and is often abundant in young plantations of Scots pine. Although good to eat, the fungus is prone to insect attack during warm weather, and for this reason the best harvests are usually made in the autumn. The pores are smaller and paler than those of *S. grevillei*. **Cap** 5–12 cm, convex, pale yellowish brown to chocolate brown, covered by a grey slime, shiny. **Tubes** adnate to decurrent, lemon yellow. **Pores** angular, pale yellow to golden, finally brownish. **Stem** 5–10 cm, solid, yellow-white with wine-coloured scales; ring slimy, white. **Flesh** thick. *July–Oct.*

Coniferous woods

## *Chroogomphus rutilus*
### Pine spike-cap

*Cap is sticky but not slimy*

*Spores 16–20 × 6–7 μm, spindle-shaped*

*Cobweb-like ring zone*

× 0.3

Also known as *Gomphidius viscidus*. Distinguished from the other two spike-caps (below) by its rich wine-red to red-brown cap which lacks a slimy veil. It has a slight scent and is edible but poor. **Cap** 4–15 cm, conical to convex with a pointed umbo, sticky, shiny when dry, smooth. **Gills** arcuate-decurrent, yellowish to purplish grey, finally purplish black, thick, forked, spaced. **Stem** 4–12 cm, cylindrical, tapers towards base, reddish at apex, saffron to chrome yellow at base; ring high, fibrillose. *Aug–Nov*.

## *Gomphidius roseus*  | ## *G. glutinosus*
### Slimy spike-cap

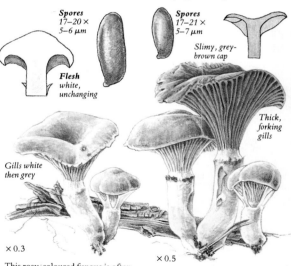

*Spores 17–20 × 5–6 μm*

*Flesh white, unchanging*

*Spores 17–21 × 5–7 μm*

*Slimy, grey-brown cap*

*Thick, forking gills*

*Gills white then grey*

× 0.3

× 0.5

This rosy-coloured fungus is often found with *Suillus bovinus* (p 40), and it is thought that the two spp may be associated. The smallest sp of the *Gomphidius* genus. **Cap** 2–5 cm, convex then depressed, sticky, smooth, shiny, pale pink to rusty orange-red. **Gills** decurrent, broad, thick, spaced. **Stem** 2–4 cm, tapers below, solid, white, covered with a pinkish sticky veil; ring fibrillose, soon disappearing. *July–Nov*.

Grows in troops near spruce. The glutinous veil should be removed before eating. **Cap** 5–14 cm, convex to depressed, often umbonate, grey-brown, darker at centre, developing black patches. **Gills** arcuate-decurrent, white to grey, finally black-brown, spaced. **Stem** 4–6 cm, tapers to yellow, slimy base, white and dry above; ring blackened by spores. *July–Nov*.

# Mixed woods

Trees that are not planted by man tend to grow in their own specialized habitats; beech and oak prefer richer soils whilst birch and pine are happier in the poor acid soils. Today most woodlands do not represent natural communities but are the result of planting over a long period and few trees are allowed to become dominant. Many woodlands therefore show a mixture of tree species, each with their own associated fungus flora.

Included in this section are those fungi which may be found under both coniferous and frondose trees. Not surprisingly, some of the commonest mushrooms of all will be found in this type of community; for example, the Blusher (*Amanita rubescens*, p 44) appears almost anywhere, while the destructive Honey fungus (*Armillaria mellea*, p 45) can grow in enormous quantities on many hosts. Several of the boletes are also very common, such as *Boletus erythropus* (p 60) and *Xerocomus chrysenteron* (p 61).

## *Amanita inaurata*

**Spores** 10–13 μm diam, spherical, non-amyloid

Cap margin is grooved

No ring on stem

Also known as *A. strangulata*. Although considered edible, be wary of all *Amanita* spp. Distinguished from other ringless forms by the loose brown fragments of veil, which remain on the cap surface, and the tall stem (up to twice the cap diameter), which is covered in wavy zones of grey scales. **Cap** 5–15 cm, convex or umbonate, yellowish to greyish brown, with strongly striate margin. **Gills** white, crowded. **Stem** 10–25 cm, white with grey, appressed scales; volva only present when young, soon reduced to fragments. *July–Sept*.

× 0.3

Mixed woods

## *Amanita rubescens*
### The blusher

*Spores 7–11 × 4–8 μm, amyloid*

*Flesh reddens when bruised*

× 0.3

A very variable sp which can cause haemolysis if eaten raw. **Cap** 5–15 cm, spherical to convex, grey to red-brown, with loose, whitish scales. **Gills** free, white spotted brown. **Stem** 5–15 cm, white above striate ring, red-brown below; scaly, swollen base lacks volva. *June–Oct.*

## *A. pantherina*
### The panther

*Spores 10–12 × 7–8 μm, non-amyloid*

*Cap with striate margin*

*Volva forms 2–3 rings*

× 0.3

Differs from *A. rubescens* in not bruising red and having a cap which is never grey, covered in pure white scales. **Cap** 5–12 cm, black-brown to olive brown then paler. **Gills** white. **Stem** 6–12 cm, white, silky, with a swollen base and a fragile, white ring. **Flesh** white, purplish brown with phenol; smells of raw potatoes. *July–Oct.*

## *A. excelsa*

*Spores 8–10 × 6–8 μm*

Also known as *A. spissa*. Mildly scented, but not recommended for eating. A variable fungus. **Cap** 6–15 cm, spherical to convex, pearl grey to greyish brown, with grey or off-white scales. **Gills** white. **Stem** 8–15 cm, robust, with greyish scaly zones and large, white, striate ring. **Flesh** firm. *June–Oct.*

*No volva but bulbous base has 3 or 4 zones of woolly scales*

× 0.3

Mixed woods

## *Armillaria mellea*
### Honey fungus

*Spores 9–10 × 5–7 μm*  *Scaly cap and stem*

Kills many trees; grows in tufts from stumps in huge numbers. Edible, but reject stems and old caps. **Cap** 5–25 cm, convex expanding, honey yellow to rust. **Gills** white to yellow, rust-spotted, crowded. **Stem** 10–20 cm, tough; ring fleshy. *June–Nov.*

*Decurrent gills*

*Spreads by black, bootlace rhizomorphs under bark*

× 0.3

## *A. tabescens*

*Spores 8–10 × 5–7 μm*

*No ring on stem*

Not as common as the Honey fungus and with a preference for frondose trees. Distinguished by the absence of a ring. **Cap** 4–12 cm, convex to campanulate, finally depressed, yellowish to reddish brown, with small, fibrillose scales. **Gills** adnate-decurrent, white to pinkish rust. **Stem** 5–12 cm, cap colour. *July–Oct.*

*Darker towards base*

× 0.3

## *Clitocybe infundibuliformis*
### Common funnel cap

*Spores 5–8 × 3–5 μm, pear-shaped*

Very common; also known as *C. gibba*. The caps are usually eaten with other fungi. Smell of bitter almond. **Cap** 3–8 cm, flesh-coloured to pale rust brown or cream, convex soon infundibuliform, dry, with thin margin. **Gills** decurrent, whitish, narrow, crowded. **Stem** 4–8 cm, solid, fibrillose. *Aug–Oct.*

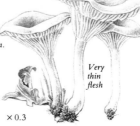

*Very thin flesh*

× 0.3

Mixed woods

## *Clitocybe brumalis*
### Winter funnel cap

*Spores 4–5 × 3–4 µm, ellipsoid*

Only occurs late in the year and usually grows in groups. The cap is dark yellowish brown with an oily appearance, drying paler. **Cap** 2–5 cm, thin, always depressed, finally infundibuliform, with a wavy margin. **Gills** decurrent, pallid, narrow, crowded. **Stem** 2–5 cm, slender, white to grey, smooth. *Nov–Dec.*

× 0.6

## *C. candicans*

*Gills are pure white*

*Spores 4–7 × 2–3 µm, ellipsoid*

Attached to dead leaves by a hairy, curving base (described as a knee-joint in France). This small, slender fungus is one of several white *Clitocybe* spp, growing either singly or in small groups. **Cap** 2–5 cm, convex to depressed, white or pale cream, smooth, dry, non-striate. **Gills** short-decurrent, white. **Stem** 3–6 cm, colour as cap, with woolly base. *Sept–Nov.*

× 0.6

## *Collybia butyracea*
### Greasy tough shank

*Cap has dark, oily centre spot*

*Spores 6–8 × 3–4 µm, ellipsoid*

The greasy cap surface lends this species its popular name. The cap pales from the margin inwards when dry, but the umbo remains coloured. Grows in clusters or rings. **Cap** 4–8 cm, convex to umbonate, bay brown to olive brown. **Gills** adnexed, white, broad, crowded. **Stem** 5–10 cm, reddish, with a swollen base, striate, tough. *Sept–Nov.*

× 0.6

Mixed woods

## *Collybia erythropus*

**Spores**
6–8 × 3–4 μm, ellipsoid

*Stem and cap contrast markedly in colour*

A difficult sp to identify; it is also known as *C. bresadolae* and *C. marasmioides*. May be distinguished from the similar *C. acervata*, which grows under spruce trees, by its less crowded gills and narrower spores. **Cap** 1–7 cm, convex to depressed, reddish brown when moist, drying to whitish. **Gills** adnate, white with a pinkish tint, narrow. **Stem** 3–7 cm, slender, rusty brown, smooth, hard, tough. *July–Nov*.

× 0.6

## *C. tuberosa*

**Spores**
3–5 × 2–3 μm, ellipsoid

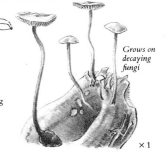

*Grows on decaying fungi*

Occurs in troops amongst moss as well as from rotting mushrooms. The stem arises from a small purplish or red-brown sclerotium. **Cap** 0.5–0.8 cm, convex expanding to flat, off-white with a yellow centre, smooth, very thin. **Gills** adnate, white, narrow, crowded. **Stem** 2–3 cm, slender, white or reddish brown. *Sept–Nov*.

× 1

## *Nyctalis asterophora*    ## *N. parasitica*

### Pick-a-back toadstool

**Spores**
5–6 × 3–4 μm

*Powdery cap*

× 1     × 1

Common in wet seasons, this species can also be seen growing on rotting species of *Lactarius* and *Russula*. **Cap** 0.5–2 cm, rounded, white, soon covered with a powdery fawn coating of star-shaped chlamydospores which tend to prevent gill development. **Stem** very short, white. *Aug–Oct*.

Grows in clusters on old *Russula* species. **Cap** 1–3 cm, campanulate expanding, white then greyish, silky. **Gills** pale, thick and spaced at first, then brownish and distorted by ellipsoid chlamydospores (15 × 10 μm). **Stem** 1–4 cm, slender, greyish. *Aug–Oct*.

Mixed woods

## *Tricholoma sejunctum*

**Spores** 6–8 × 4–6 μm, ellipsoid

Easily confused with the Death cap (p 104) because of its greenish-yellow cap, but it lacks a ring and volva on the stem. The flesh smells of flour and has a bitter, unpleasant taste, even after cooking. The cap margin is typically paler than the rest of the cap. Occurs either singly or in groups. **Cap** 4–12 cm, campanulate to convex, greenish yellow with brown centre, radially fibrillose, shiny, slimy when moist. **Gills** sinuate, white discolouring yellow. **Stem** 4–8 cm, solid, white often tinted yellow. *Sept–Nov*.

× 0.3

## *T. virgatum*

**Spores** 6–8 × 5–6 μm, ellipsoid

The shiny ash-grey cap is distinctly pointed with dark fibrils radiating from the centre. The flesh is greyish white and has a damp, earthy smell. **Cap** 5–7 cm, conical to campanulate, expanding. **Gills** sinuate, white, crowded. **Stem** 5–9 cm, tall, white, faintly fibrillose. *Oct–Nov*.

*Gills have jagged edge*

*Sharp, bitter taste*

× 0.5

## *Tricholomopsis decora*

**Spores** 5–7 × 4–5 μm

Grows in troops on roots and trunks of rotting woods, especially of conifers, in mountainous areas. Similar to *T. rutilans* (p 17), but it is covered in brown, not purplish, scales. **Cap** 4–10 cm, convex then depressed with a wavy margin, golden yellow with numerous minute, brown, appressed scales, dry. **Gills** sinuate, golden yellow, very broad, crowded. **Stem** 4–8 cm, cylindrical, same colour as cap, fibrillose-scaly, often curved. *July–Oct*.

× 0.5

Mixed woods

## *Micromphale foetidum*

 **Spores** $8–10 \times 3–4 \mu m$, ellipsoid

Furrows radiate from cap centre

The unpleasant smell of this fungus resembles rotting fish. May be found in troops on fallen branches and dead twigs, especially beech and hazel. **Cap** 2–4 cm, convex expanding, finally with a small central depression, thin, reddish brown to date brown. **Gills** slightly decurrent, pinkish yellow, spaced, interveined. **Stem** 2–4 cm, short, date brown to black, hollow, finely velvety. *Aug–Nov.*

Stem narrows near base

× 0.6

## *Xeromphalina cauticinalis*

 **Spores** $5–7 \times 3–4 \mu m$, ellipsoid

Grows in clusters on logs and stumps, sometimes amongst pine litter. **Cap** 1–3 cm, convex with a small central depression, rusty yellow, darker at centre, deeply grooved. **Gills** adnexed, cream. **Stem** 2–4 cm, thin, yellow above, tawny to black below, shiny, with a swollen base which is covered in stiff, red-brown hairs. *Sept–Oct.*

Long stem blackens towards base

× 1

## *Fayodia bisporigera*

 **Spores** $8–11 \mu m$ diam, spherical, amyloid, ornamented

A fairly rare species in England, mainly confined to Scotland, but more common in central and northern Europe. *Fayodia* is a small genus resembling *Mycena* and *Omphalina* but with unusually ornamented spores. *F. bisporigera* tends to grow amongst dead twigs on rich soil. **Cap** 2–3 cm, convex to depressed, grey-brown drying to pale brown, with a striate margin. **Gills** adnate, grey, broad, spaced. **Stem** 2–4 cm, slender, pale grey, hollow, smooth. *Aug–Sept.*

Stem paler than cap

× 0.6

Mixed woods

## Mycena galopus

### Milk-drop mycena

**Spores** 10–13 × 5–6 μm

Stem exudes copious white milk if broken

×1

Grows amongst dead leaves, never on living trees. **Cap** 0.5–1.5 cm, conical to campanulate, greyish to pale brown, darker centre, striate. **Gills** adnexed, white. **Stem** 6–10 cm, paler at apex. *July–Oct.*

## M. metata

**Spores** 9–10 × 4–5 μm, amyloid

Strong smell of iodoform ▶

×1

Recognized by its opaque, flesh-coloured cap and distinctive smell. **Cap** 1–2 cm, conical, umbonate, greyish pink, often brownish in centre, smooth. **Gills** adnate, white, fairly spaced. **Stem** 5–8 cm, thin, grey-white. *Aug–Nov.*

## M. rorida

Stem has slimy coat

**Spores** 12–13 × 4–7 μm, amyloid

In troops on fallen wood

×1

In dry weather, when the stem is not slimy, it may be mistaken for *Marasmiellus ramealis* (p 69). **Cap** 0.5–1 cm, whitish or pale yellow, dry, grooved. **Gills** white, spaced. **Stem** 2–3 cm, white. *Aug–Nov.*

## Lepiota clypeolaria

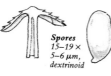

**Spores** 15–19 × 5–6 μm, dextrinoid

Cap surface breaks up into felty patches

A fairly large species of *Lepiota* with several varieties based on differences in scaliness of stem. A similar but more robust species, *L. ventriosospora*, may be distinguished by the yellowish to pinkish-brown ring on its stem. **Cap** 2–8 cm, campanulate expanding, covered with a yellowish-brown, velvety layer which soon cracks into scales, with a shaggy margin. **Gills** free, creamy white, crowded. **Stem** 6–8 cm, colour as cap, scaly below thin, ephemeral ring. *Aug–Sept.*

×0.5

Mixed woods

## *Lepista nuda*

### Wood blewit

**Spores** 6–8 × 4–5 µm, ellipsoid, roughened with minute warts

**Flesh** white with a lilac tint; varies from thin to thick in texture

The lilac-blue colour of this excellent edible mushroom must make it one of the best-known species in northern Europe. However, it can be slightly toxic if eaten raw (some people are allergic to it) and therefore par-boiling is recommended. Occurs in groups, often forming rings on soil that has been disturbed. Sold in country markets. **Cap** 7–12 cm, convex to depressed with wavy margin, moist, lilac-grey. **Gills** sinuate, bright lilac but soon discolour. **Stem** 4–8 cm, bright violet, fibrous. *Sept–Dec*.

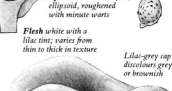

Lilac-grey cap discolours grey or brownish

Stem may be slender or robust

Crowded gills

× 0.5

## *Russula ochroleuca*

### Common yellow russula

An exceptionally common fungus, often one of the first to appear in late summer. Recognized by the ochre-yellow cap, pale cream gills and greyish stem. The taste is variable, ranging from very hot to almost mild. Other yellow species with which it may be confused include: *R. claroflava* (p 99), which is more brightly coloured with yellow gills; *R. fellea* (p 113), which is uniformly straw-coloured; and *R. lutea*, which is smaller with egg-yellow gills. **Cap** 4–10 cm, convex to depressed, yellow but often with shades of ochre, orange, brown or grey-green. **Gills** quite crowded. **Stem** 3–8 cm, firm, cylindric or slightly swollen at base. *Aug–Nov*.

**Spores** 8–10 × 7–8 µm, amyloid

Gills are pale cream

White stem turns greyish with age

× 0.5

Ochre-yellow cap has slightly grooved margin

Mixed woods

## *Russula delica*

### Milk-white russula

*Spores 8–12 × 7–9 μm, warty*

Of poor quality: grill and season. Resembles a *Lactarius* sp, but does not produce a milk. **Cap** 8–15 cm, whitish. **Gills** decurrent, thick, pale, spotted brown. **Stem** 3–5 cm, short, hard, white. *June–Oct.*

*Firm texture*

*Slightly fishy smell*

× 0.3

## *R. nigricans*

### Blackening russula

*Spores 7–8 × 6–7 μm*

The largest *Russula* sp. Edible but poor. Often attacked by *Nyctalis* spp when old. Flesh turns from red to black on cutting. **Cap** 10–20 cm, white to brown, finally black. **Gills** adnate, whitish. **Stem** 3–7 cm, thick. *July–Nov.*

*Gills thick and widely spaced*

*All parts blacken with age*

× 0.25

## *R. foetens*

### Foetid russula

*Spores 8–10 × 7–9 μm, strongly warty, amyloid*

Recognized by its disgusting, rancid smell and slimy, grooved cap. *R. laurocerasi* is similar but less common and has a pleasant smell of marzipan. Mature specimens of *R. foetens* are often infested by insect larvae. **Cap** 10–18 cm, convex to flat, splitting, pale ochre to tawny. **Gills** whitish spotted brown, broad, crowded. **Stem** 10–13 cm, dirty white. **Flesh** white, bruises red. *June–Oct.*

*Cap is very slimy*

× 0.3

52

Mixed woods

## R. fragilis
### Fragile russula

*Spores 7–9 × 6–8 μm, with network, amyloid*

Smaller, more fragile, with a cap colour not as bright as *R. emetica* (p 25). Smells of apples and has an acrid taste. **Cap** 2–6 cm, convex to depressed, sticky, pinkish grey to violet, often with grey-green tints, centre darker; margin grooved. **Gills** serrated. **Stem** 2–5 cm, soft. *July–Oct.*

*Gills, flesh and stem are pure white*

× 0.6

## R. puellaris

*Spores 6–9 × 5–7 μm, warty, amyloid*

A small, delicate species with all parts yellowing with age. It has a mild taste and is edible, although not very fleshy. **Cap** 3–6 cm, convex to depressed, sticky, colour varying from pink to light purple, with shades of yellow and green; margin ridged, thin. **Gills** creamy yellow, fairly crowded. **Stem** 3–6 cm, white, fragile. *July–Oct.*

*Fragile, becoming more yellow*

× 0.6

## R. xerampelina

*Cap is very variable in colour*

*Spores 8–11 × 6–9 μm*

Considered one of the best edible *Russula* spp, but the characteristic smell of crab will persist even in cooking. The gills, flesh and stem generally discolour yellowish brown. **Cap** 5–12 cm, rounded to slightly depressed, dry, sticky when moist, bright purplish red, blackish at centre, often with brownish tints. **Gills** whitish to pale ochre. **Stem** 3–11 cm, white with wine-red tints. **Flesh** white, stains green with ferrous sulphate (not known for any other *Russula*). *Aug–Nov.*

*Smell of crab*

× 0.5

Mixed woods

## Lactarius torminosus
### Woolly milk cap

**Spores**
7–10 ×
6–7 µm,
amyloid

*Felty cap with concentric salmon-pink zones*

*Shaggy margin*

Recognized by the zoned woolly cap from which exudes a copious hot white milk. **Cap** 4–15 cm, convex then depressed with margin remaining incurved; pale pink, slimy when moist, covered with long, dense, cottony, pink-red hairs. **Gills** decurrent, thin, crowded, pinkish cream. **Stem** 3–8 cm, stocky, soon hollow, creamy, with pink zone near apex. *Aug–Nov.*

*Likes damp places*

× 0.5

## L. trivialis

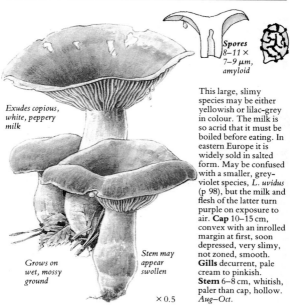

**Spores**
8–11 ×
7–9 µm,
amyloid

*Exudes copious, white, peppery milk*

This large, slimy species may be either yellowish or lilac-grey in colour. The milk is so acrid that it must be boiled before eating. In eastern Europe it is widely sold in salted form. May be confused with a smaller, grey-violet species, *L. uvidus* (p 98), but the milk and flesh of the latter turn purple on exposure to air. **Cap** 10–15 cm, convex with an inrolled margin at first, soon depressed, very slimy, not zoned, smooth. **Gills** decurrent, pale cream to pinkish. **Stem** 6–8 cm, whitish, paler than cap, hollow. *Aug–Oct.*

*Grows on wet, mossy ground*

*Stem may appear swollen*

× 0.5

Mixed woods

## *L. serifluus*

### Watery milk cap

*Spores 8–9 × 7–8 µm, network, amyloid*

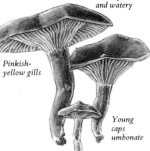

Milk sparse and watery

Pinkish-yellow gills

Young caps umbonate

A small fungus occurring in troops on sandy soil. It produces a scant, watery milk which has a mild taste and a rather oily smell. **Cap** 2–6 cm, umbonate soon depressed, wavy, deep rusty brown, darker at centre, drying paler. **Gills** decurrent, pinkish yellow or reddish, crowded. **Stem** 3–5 cm, colour as cap, tapers below. *July–Oct.*

× 0.6

## *L. mitissimus*

*Spores 8–10 × 6–8 µm, amyloid*

Cap smooth, not zoned

Taste mild, then bitter

The bright orange-brown cap resembles *L. volemus* (p 114), but it is much smaller and associated with hazel or oak. **Cap** 2–6 cm, convex to depressed with small umbo; margin thin. **Gills** slightly decurrent, deep ochre yellow, crowded. **Stem** 4–8 cm, colour as cap, whitish base. *Aug–Oct.*

× 0.6

## *L. fuliginosus*

*Spores 8–9 × 7–8 µm, amyloid*

The white flesh of this milk cap reddens rapidly on exposure to air, but the milk, which is white with a slightly acrid taste, turns saffron yellow. **Cap** 2–10 cm, convex to depressed, pale to dark sooty brown, dry, shortly velvety then smooth. **Gills** scarcely decurrent, cream to ochre yellow, bruising red, spaced. **Stem** 6–8 cm, white tinted brown, tapering below. *Aug–Oct.*

Velvety cap

× 0.5

Mixed woods

## *Cortinarius alboviolaceus*

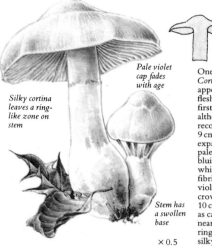

**Spores** 8–10 × 5–6 μm, ellipsoid

*Pale violet cap fades with age*

*Silky cortina leaves a ring-like zone on stem*

*Stem has a swollen base*

× 0.5

One of the commoner *Cortinarius* spp, often appearing early. The flesh, which is bluish at first, is probably edible although not recommended. **Cap** 3–9 cm, convex expanding, umbonate, pale blue-violet to bluish white, becoming whitish, silky-fibrillose. **Gills** pale violet to clay brown, crowded. **Stem** 5–10 cm, robust, colour as cap, deeper violet near apex, with distinct ring zone left by white, silky veil. *Aug–Nov*.

## *C. caerulescens*

**Spores** 9–11 × 5–6 μm, almond-shaped

*Cap centre discolours yellowish brown*

× 0.6

Recognized by its dark bluish sticky cap and deep blue gills which become discoloured rusty brown by the spores. A similar sp, *C. caesiocyaneus* (p 117), has a paler cap and the gills are whitish at first. *C. caerulescens* occurs on chalky soil. **Cap** 5–10 cm, convex, violet blue with discoloured centre. **Gills** crowded. **Stem** 5–8 cm, violet blue with ochre-yellow, club-shaped, marginate base; lilac cortina visible. *Aug–Nov*.

*Distinct rim around stem base*

Mixed woods

## *C. cyanites*

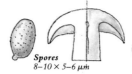

**Spores**
$8–10 \times 5–6\ \mu m$

More common in Scotland and in mountainous areas, especially under pine or birch. **Cap** 5–8 cm, pale blue with a carmine-red centre, sticky when young, later dry, fibrillose. **Gills** deep bluish grey then rusty brown, crowded. **Stem** 6–8 cm, bluish, bruising red, with the yellowish veil forming conspicuous scales on the lower part. **Flesh** turns reddish when cut. *Aug–Nov*.

× 0.5

## *C. tabularis*

**Spores**
$7–9 \times 5–7\ \mu m$, *ellipsoid*

*Lower part of stem is yellowish*

× 0.5

*C. azureus* is closely related but has a violet cap and stem. *C. tabularis* often occurs under beech. **Cap** 3–8 cm, convex with an umbo, pale yellowish brown, darker at centre, shiny, smooth, sticky when young. **Gills** clay brown to rusty brown, crowded. **Stem** 3–12 cm, tall, slightly swollen at the base, whitish, pale yellow below, with remnants of whitish cortina. *Aug–Nov*.

## *Inocybe lacera*

**Spores**
$10–14 \times 4–6\ \mu m$, *cylindric*

One of a small group of *Inocybe* spp with exceptionally long, cylindric spores. Several related spp are to be found in sand dunes. *I. lacera* is frequently found by the sides of paths through woods, especially on sandy soil. **Cap** 2–4 cm, convex with small umbo, brown, fibrillose-scaly. **Gills** adnexed, olive grey to cigar brown, broad. **Stem** 3–4 cm, slender, tapering towards base, colour as cap, fibrillose. **Flesh** rusty brown in stem base. *June–Nov*.

*Scaly cap*

*Brown, fibrillose stem*

× 1

Mixed woods

## *Hypholoma fasciculare*
### Sulphur tuft

*Young gills greenish*

**Spores** 6–7 × 3–5 µm, with germ-pore

Occurs in large clusters on stumps. Has a bitter taste and acts as a gastro-enteric irritant. **Cap** 2–5 cm, pale yellow, brown in centre, smooth. **Gills** sulphur yellow, finally purplish black, crowded. **Stem** 5–15 cm, colour as cap, rusty at base, curved, fibrillose. *Jan–Dec.*

× 0.5

## *H. sublateritium*
### Brick-red hypholoma

*Brick-red cap*

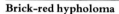

**Spores** 6–7 × 3–4 µm, with germ-pore

One of the largest species of *Hypholoma*; more robust but less common than the Sulphur tuft. Grows in clusters on old stumps and is usually identified by its brick-red cap and stem. **Cap** 4–10 cm, convex, fleshy, red-brown, yellowish towards margin, smooth. **Gills** adnate, yellowish to very dark brown, crowded. **Stem** 5–10 cm, yellowish brown, red-brown below, fibrillose-scaly. *Aug–Nov.*

*Greenish-yellow gills*

*Robust, scaly stem*

× 0.3

## *Pholiota lenta*

*Sticky surface*

**Spores** 6–8 × 3–4 µm, with germ-pore

Used to be placed in the genus *Flammula*, along with other spp of *Pholiota* without a membranous ring on the stem. The sticky pale cap and yellowish-brown gills are characteristic. **Cap** 5–12 cm, pale yellowish, darker in centre, with small, white scales when young. **Gills** slightly decurrent, yellowish brown to rusty brown, crowded. **Stem** 6–8 cm, whitish, brownish near base, fibrillose. *Oct–Nov.*

× 0.3

*Fruity smell*

58

Mixed woods

## *Agaricus augustus*

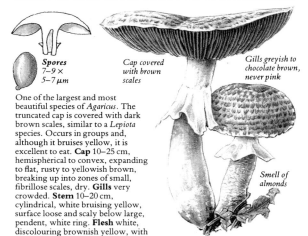

**Spores** 7–9 × 5–7 μm

*Cap covered with brown scales*

*Gills greyish to chocolate brown, never pink*

*Smell of almonds*

One of the largest and most beautiful species of *Agaricus*. The truncated cap is covered with dark brown scales, similar to a *Lepiota* species. Occurs in groups and, although it bruises yellow, it is excellent to eat. **Cap** 10–25 cm, hemispherical to convex, expanding to flat, rusty to yellowish brown, breaking up into zones of small, fibrillose scales, dry. **Gills** very crowded. **Stem** 10–20 cm, cylindrical, white bruising yellow, surface loose and scaly below large, pendent, white ring. **Flesh** white, discolouring brownish yellow, with a smell of bitter almonds. *July–Oct.*

× 0.25

## *A. macrosporus*

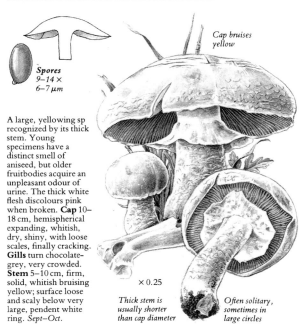

**Spores** 9–14 × 6–7 μm

*Cap bruises yellow*

A large, yellowing sp recognized by its thick stem. Young specimens have a distinct smell of aniseed, but older fruitbodies acquire an unpleasant odour of urine. The thick white flesh discolours pink when broken. **Cap** 10–18 cm, hemispherical expanding, whitish, dry, shiny, with loose scales, finally cracking. **Gills** turn chocolate-grey, very crowded. **Stem** 5–10 cm, firm, solid, whitish bruising yellow; surface loose and scaly below very large, pendent white ring. *Sept–Oct.*

× 0.25

*Thick stem is usually shorter than cap diameter*

*Often solitary, sometimes in large circles*

59

Mixed woods

## *Boletus calopus*

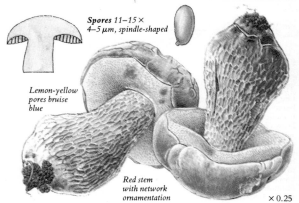

*Spores 11–15 × 4–5 μm, spindle-shaped*

*Lemon-yellow pores bruise blue*

*Red stem with network ornamentation*

× 0.25

One of the most beautifully coloured mushrooms, this sp might be confused with *B. luridus* (p 122) or *B. erythropus* (below), both of which have red pores. The thick, veined stem is often swollen and the large, fleshy cap is generally convex, although the size and shape may vary considerably. The creamy-white flesh turns blue when broken and is too bitter to be of any culinary value. Occurs more frequently in mountainous areas on acid soils. **Cap** 5–15 cm, convex expanding, whitish grey to pale brown, dry, velvety, at times cracked. **Tubes** lemon yellow, bruise blue. **Pores** round. **Stem** 6–10 cm, solid, sulphur yellow above, purplish red below. *Aug–Oct*.

## *B. erythropus*

*Cap dark brown with olive or reddish tints*

*Spores 12–16 × 4–6 μm, spindle-shaped*

The firm, bright yellow flesh discolours first blue then green when broken open, and for this reason it is rarely regarded as an edible species. However, it is excellent when properly cooked, with a mild to slightly acid taste, although it can cause vomiting if eaten raw. Do not confuse this bolete with the poisonous *B. satanas* (p 122), which always has a pale cap and white flesh. **Cap** 8–20 cm, convex, dry, velvety, then smooth, shiny. **Tubes** free, olive brown staining blue. **Pores** small, round, yellow to orange-red, turning blue when touched. **Stem** 7–15 cm, robust, dotted orange-yellow above, brown below, ridged. *May–Nov*.

*No network on stem*

× 0.25

## *Xerocomus chrysenteron*
### Red-cracked boletus

## *X. subtomentosus*
### Yellow-cracked boletus

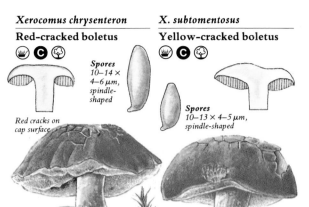

Edible but only mediocre, this is the most common British bolete. The soft, creamy flesh blues very slowly on exposure. **Cap** 4–8 cm, convex to depressed, dry, velvety, often cracking, pale olive to reddish brown. **Tubes** adnate to decurrent, yellow then greenish. **Pores** large, angular, yellow, bruising blue. **Stem** 4–8 cm, slender, solid, reddish, fibrillose. *June–Nov.*

Very similar to *X. chrysenteron* but lacks red on cap and stem. A good edible species: the young; firm caps are worth pickling in vinegar. **Cap** 4–8 cm, convex expanding, dry, velvety, finally cracking, pale olive yellow to rusty brown (not blood red). **Tubes** olive yellow. **Pores** large, golden yellow, sometimes bluing if bruised. **Stem** 3–9 cm, striate, yellowish brown. **Flesh** whitish to pale yellow. *June–Nov.*

## *Chalciporus piperatus*
### Peppery boletus

Common on sandy soil in birch-pine woods. Not usually eaten because of peppery taste, but has been used as a condiment when dried. Recognized by the tapering stem with yellow base and cinnamon pores. **Cap** 3–12 cm, convex to flat, uniformly dull brown, sticky when wet. **Tubes** adnate to decurrent. **Pores** small, angular. **Stem** 4–10 cm, solid, reddish brown. **Flesh** yellowish, thick. *Aug–Nov.*

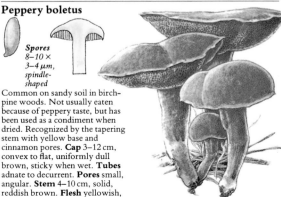

# Frondose woods

A frondose wood consists of broad-leafed trees only, without any conifers; it often embraces many kinds of trees, and generally offers a wide choice of mushrooms. Many of the fungi grouped in this habitat are also found in beech, oak and birch woods, but they are included here because they grow under other broad-leafed trees as well, and are not restricted to a specific tree. At the same time, trees such as ash, hazel, poplar and lime have their own distinctive species, and knowledge of these associations can be an important guide to identification.

## *Amanita solitaria*

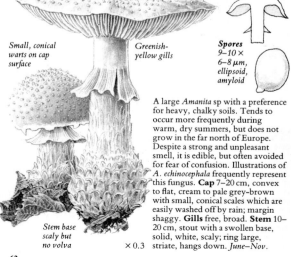

*Small, conical warts on cap surface*

*Greenish-yellow gills*

**Spores** 9–10 × 6–8 µm, ellipsoid, amyloid

*Stem base scaly but no volva*

× 0.3

A large *Amanita* sp with a preference for heavy, chalky soils. Tends to occur more frequently during warm, dry summers, but does not grow in the far north of Europe. Despite a strong and unpleasant smell, it is edible, but often avoided for fear of confusion. Illustrations of *A. echinocephala* frequently represent this fungus. **Cap** 7–20 cm, convex to flat, cream to pale grey-brown with small, conical scales which are easily washed off by rain; margin shaggy. **Gills** free, broad. **Stem** 10–20 cm, stout with a swollen base, solid, white, scaly; ring large, striate, hangs down. *June–Nov*.

## *Hygrophorus dichrous*

## *Clitocybe fragrans*

Edible and good, but must be well cooked. **Cap** 4–8 cm, conical then umbonate with an inrolled margin, greyish to olive brown, darker centre, shiny. **Gills** decurrent, spaced. **Stem** 5–9 cm, solid, covered with bands of olive-brown scales. *Aug–Oct*.

This aniseed-scented sp differs from other similar-smelling spp by its pale cap and relatively long stem. **Cap** 2–5 cm, convex to slightly depressed, pale yellow with a brown centre when moist. **Gills** slightly decurrent, whitish, crowded. **Stem** 5–8 cm, cap colour, smooth. *Aug–Dec*.

## *C. odora*
### Blue-green clitocybe

## *C. suaveolens*

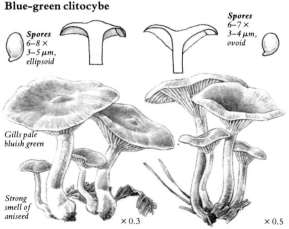

Has been used as a condiment for its taste and smell of aniseed. A variable sp in size and colour; typically the cap is uniformly blue-green, but discolours greyish as it dries. **Cap** 3–10 cm, convex, never depressed. **Gills** adnate to short-decurrent. **Stem** 4–10 cm, curved, cream with a pinkish tint. **Flesh** thin, whitish. *Aug–Oct*.

A species extremely closely related to *C. fragrans* (above), and thought by some not to be distinct. It differs in having a depressed cap with a striate margin and dark brown centre. Has a fragrant smell of aniseed. **Cap** 3–6 cm, whitish, depressed. **Gills** decurrent, narrow, whitish. **Stem** 5–7 cm, whitish. *Sept–Nov*.

Frondose woods

## *Clitocybe dicolor*

**Spores** $5–7 \times 3–4\ \mu m$, ellipsoid

*Grey-brown when moist, drying white*

Variable, with the cap dark greyish brown when water-soaked but drying to almost white. One of a group of small *Clitocybe* spp with grey gills, but differs in lacking a mealy smell. **Cap** 2–6 cm, convex, umbilicate, smooth. **Gills** decurrent, grey. **Stem** 3–6 cm, slender, hollow, whitish at apex, greyish brown below. *Sept–Nov.*

$\times 0.5$

## *C. geotropa*

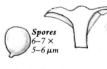

**Spores** $6–7 \times 5–6\ \mu m$

Likes chalky soil in wood clearings. The large cap always has an umbo and is edible when young, but soon becomes too tough. Often forms fairy rings; in one case estimated to have existed for over 700 years. **Cap** 10–20 cm, convex or depressed, umbonate, with inrolled margin, dry, brownish yellow, whitish when old. **Gills** decurrent, crowded, whitish. **Stem** 6–15 cm, cap colour. **Flesh** firm, white. *Sept–Nov.*

*A large robust sp with a tall cylindric stem*

$\times 0.25$

## *C. inornata*

**Spores** $8–10 \times 3–4\ \mu m$, spindle-shaped

*Cap has inrolled, grooved margin*

May be confused with *Paxillus involutus* (p 101), but much rarer. It grows either singly or in small groups in moist woods, although in North America it is more likely to be found in conifer plantations. It has a persistent smell of caraway seed or radish. **Cap** 4–10 cm, soon flat, pale brown to greyish yellow, powdery. **Gills** decurrent, greyish brown, broad. **Stem** 2–5 cm, greyish, solid. *Sept–Nov.*

*On chalky soil, in grassy places*

$\times 0.5$

## *Pseudoclitocybe cyathiformis*  *P. expallens*

### The goblet

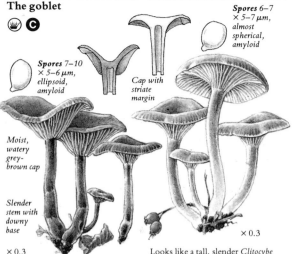

**Spores** 7–10 × 5–6 μm, ellipsoid, amyloid

*Moist, watery grey-brown cap*

*Slender stem with downy base*

× 0.3

**Spores** 6–7 × 5–7 μm, almost spherical, amyloid

*Cap with striate margin*

× 0.3

Appears in late summer, continuing through until the first frosts. Grows amongst woody debris in wet areas. **Cap** 2–6 cm, dark grey-brown drying paler, infundibuliform. **Gills** adnate to decurrent, smoky grey, spaced. **Stem** 5–10 cm, cap colour or paler, fibrillose. **Flesh** thin, watery. *Sept–Nov*.

Looks like a tall, slender *Clitocybe* sp, but more closely related to the genus *Melanoleuca*. Difficult to separate from *P. cyathiformis*; the cap margin is more grooved and the fruitbodies tend to be smaller and paler. **Cap** 2–5 cm, depressed, smoky grey, faintly zoned. **Gills** decurrent, spaced, pale grey. **Stem** 4–8 cm, cap colour or paler, tough, smooth. *Sept–Nov*.

## *Lepista irina*

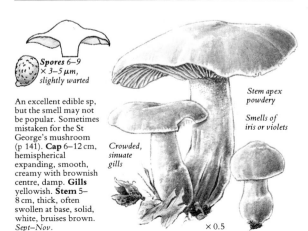

**Spores** 6–9 × 3–5 μm, slightly warted

An excellent edible sp, but the smell may not be popular. Sometimes mistaken for the St George's mushroom (p 141). **Cap** 6–12 cm, hemispherical expanding, smooth, creamy with brownish centre, damp. **Gills** yellowish. **Stem** 5–8 cm, thick, often swollen at base, solid, white, bruises brown. *Sept–Nov*.

*Crowded, sinuate gills*

*Stem apex powdery*

*Smells of iris or violets*

× 0.5

## *Laccaria amethystea*
### Amethyst deceiver

## *L. laccata*
### The deceiver

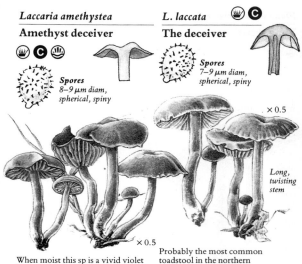

**Spores** 8–9 μm diam, spherical, spiny

**Spores** 7–9 μm diam, spherical, spiny

× 0.5

Long, twisting stem

× 0.5

When moist this sp is a vivid violet blue, but old, dry specimens may be pale lilac or even whitish. Edible but tasteless. **Cap** 1–6 cm, convex soon depressed, finely scaly, often split. **Gills** adnate-decurrent, spaced. **Stem** 4–10 cm, curved, cap colour. **Flesh** thin, pale lilac. *Aug–Nov*.

Probably the most common toadstool in the northern hemisphere but also one of the most variable, hence the name. Edible but poor. **Cap** 1–5 cm, convex to depressed, irregular, pinkish to yellowish brown, scaly. **Gills** pinkish, powdery, spaced. **Stem** 5–8 cm, colour as cap. *Aug–Nov*.

## *Cantharellus lutescens*

## *Collybia peronata*
### Wood woolly-foot

**Spores** 10–11 × 7–8 μm

**Spores** 7–9 × 3–4 μm, ellipsoid

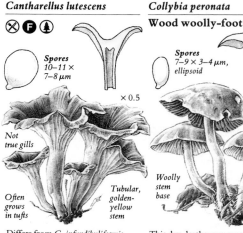

× 0.5

Not true gills

Often grows in tufts

Tubular, golden-yellow stem

× 0.5

Woolly stem base

Differs from *C. infundibuliformis* (p 105) in its orange-yellow, vein-like gills and golden stem. More delicious than Chanterelle (p 103). **Cap** 2–6 cm, funnel-shaped, wavy, grey-brown. **Gills** low, branching. **Stem** 3–8 cm, hollow. *Sept–Oct*.

This dry, leathery sp grows in small tufts amongst leaf litter. The flesh is yellowish with a hot, acrid taste. **Cap** 3–6 cm, convex, red-brown drying paler. **Gills** adnexed, yellowish, spaced. **Stem** 5–9 cm, tapers above, yellowish. *Aug–Nov*.

Frondose woods

## *Flammulina velutipes*
### Velvet shank

**Spores** $8–10 \times 4–5 \, \mu m$

Sticky, shiny cap

Dark, velvety stem

Grows tufted on wood, as a wound parasite ×0.6

A widespread fungus, suitable for soups; of value because it occurs when few other spp are available. Only the caps are edible, but remove the slimy pellicle. **Cap** 3–6 cm, convex to depressed, honey yellow, spotting brown. **Gills** adnexed, yellowish. **Stem** 4–10 cm, tough, black-brown, pale above. *Sept–Mar.*

## *Mycena haematopus*
### Bleeding mycena

**Spores** $7–10 \times 5–6 \, \mu m$, ellipsoid, amyloid

Blood-red juice in stem

×0.5

Forms small tufts on stumps; easily recognized by the blood-red juice exuded when the stem is broken. **Cap** 2–3 cm, campanulate, deep red-brown, slightly striate. **Gills** adnate, white to pinkish. **Stem** 4–7 cm, slender, cap colour. *Aug–Nov.*

## *M. tenerrima*   | ## *Collybia cirrhata*

**Spores** $8–10 \times 5–7 \, \mu m$, ovoid, amyloid

Minute, delicate fruit-bodies

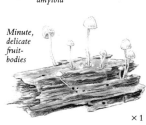
×1

This minute toadstool is often overlooked, but commonly occurs in large numbers over woody debris. **Cap** 2–6 mm, convex, very thin, white, striate, with tiny, granular scales. **Gills** adnexed, white, crowded. **Stem** 1–3 cm, very thin, white-translucent. *Aug–Nov.*

**Spores** $5–6 \times 2–3 \, \mu m$, ovoid

Silky white cap

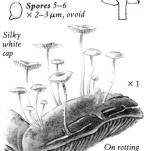
×1

On rotting fungi

Very similar to *C. tuberosa* (p 47) but does not grow from a sclerotium. **Cap** 2–12 mm, convex then flat, whitish with pale brown centre, dry, striate. **Gills** adnate, white, narrow, crowded. **Stem** 1–2.5 cm, thin, cap colour or yellowish, often curved, deeply rooting. *July–Oct.*

## *Mycena corticola*  ## *M. olida*

**Spores** 7–11 × 7–10 μm

× 1

Pink to bluish-grey caps

**Spores** 6–9 × 4–6 μm

White cap with yellow centre

In clusters on stumps

A small, delicate sp found on various mossy tree trunks in late autumn. **Cap** 5–10 cm, hemispherical then centrally depressed, radially grooved, translucent. **Gills** adnate, grey, paler at edge. **Stem** 1–2 cm, thin, paler than cap, powdery. *Oct–Dec*.

× 1

Similar in appearance to *M. flavoalba* (p 148) but with a velvety stem. **Cap** 0.5–2 cm, conical, white tinted pale yellow at centre, radially grooved. **Gills** adnate, white, crowded. **Stem** 1–3 cm, slender, white. *Aug–Nov*.

## *M. galericulata*
### Bonnet mycena

**Spores** 9–12 × 6–8 μm

## *M. polygramma*
### Steely-stemmed mycena

**Spores** 9–11 × 6–7 μm

## *M. crocata*

**Spores** 7–10 × 5–6 μm

Bruises saffron

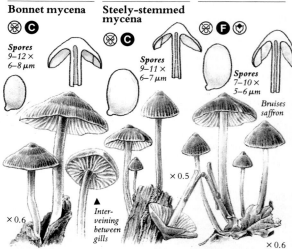

× 0.6

Inter-veining between gills

× 0.5

× 0.6

A robust *Mycena* sp; occurs in clusters on stumps, with stems often sticking together. Faint, rancid odour. **Cap** 2–6 cm, conical to umbonate, grey-brown to yellowish, almost white, grooved. **Gills** slightly decurrent, white to pink. **Stem** 3–8 cm, tough, smooth, shiny grey. *May–Nov*.

Solitary or gregarious, on or around stumps. Characterized by its silvery-grey, finely grooved stem. **Cap** 2–5 cm, conical or campanulate, slightly umbonate, grey-brown, striate at margin. **Gills** white or pale pink, quite spaced. **Stem** 5–10 cm, slender, rooting. *Aug–Nov*.

Not common in Britain; more abundant farther north, where it is often found under beech. The stem releases a carrot-red juice when broken. **Cap** 1–3 cm, conical, olive-grey. **Gills** adnexed, white staining saffron, quite spaced. **Stem** 7–10 cm, saffron, paler at apex. *Aug–Nov*.

Frondose woods

## *Gerronema chrysophyllum*

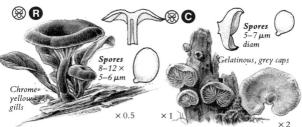

Like a large *Omphalina* sp, but has bright yellow gills. **Cap** 2–5 cm, convex with deep depression, yellow-brown, dark centre, moist. **Gills** spaced. **Stem** 1–3 cm, always short, curved. *Sept–Nov*.

## *Resupinatus applicatus*

A tiny sp, often overlooked. **Cap** 0.5–1 cm, cup- or kidney-shaped, translucent, striate, covered by white powder. **Gills** radiate from point of attachment, narrow, crowded, grey. *Aug–Nov*.

## *Marasmiellus ramealis*

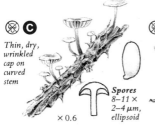

Often on dead bramble stems. Old specimens discolour pale yellowish. **Cap** 0.5–1 cm, convex then flat, thin, off-white, pinkish at centre. **Gills** few, adnate, white, narrow, spaced. **Stem** 5–10 cm, whitish or tinted reddish below. *June–Oct*.

## *Marasmius epiphyllus*

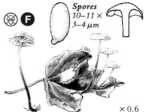

Differs from *M. androsaceus* (p 18) in having fewer gills and a paler stem. **Cap** 0.5–1 cm, thin, flat, finally centrally depressed, milk white. **Gills** interveined, narrow, white. **Stem** 1–3 cm, white above, red-brown below, velvety. *Aug–Nov*.

## *Marasmius rotula*
## Little wheel toadstool

Another small, wiry toadstool found on wood fragments and fallen branches. Sometimes fine black rhizomorphs, resembling horse hairs, are associated with the fruitbodies. The common name is derived from the gills being attached to a circular collar around the top of the stem. This connecting collar distinguishes it from *M. epiphyllus* (above). **Cap** 0.5–1 cm, convex expanding with small central depression, thin, whitish, with radial grooves. **Gills** white, broad, very spaced, equal. **Stem** 2–4 cm, black, shiny, wiry. *Aug–Nov*.

## *Tricholoma saponaceum*
### Soap-scented tricholoma

**Spores** 5–6 × 3–4 μm

*Cap surface cracking when dry*

*Soapy smell*

× 0.3

Very variable, but recognized by the distinctive odour of "mild laundry soap". The greenish tints could lead to confusion with either Death cap (p 104) or *T. sejunctum* (p 48). Can cause severe indigestion. **Cap** 5–10 cm, convex expanding, irregular, dull grey-brown, often tinted olive, smooth. **Gills** sinuate, cream tinted pink, spaced. **Stem** 5–10 cm, thick, tapers below, whitish, fibrillose. **Flesh** firm, white, discolours red. *July–Oct.*

## *T. sulphureum*
### Sulphurous tricholoma

**Spores** 8–14 × 5–6 μm, almond-shaped

*Thick, spaced gills*

× 0.5

An entirely sulphur-yellow mushroom which is locally common, preferring stiff clay soils. It is poisonous with a bitter taste and a strong, unpleasant, sweetish smell of coal gas. **Cap** 4–8 cm, fleshy, convex, slightly umbonate, irregular, dry, silky-smooth. **Gills** sinuate-adnexed, yellow. **Stem** 5–10 cm, slender, cylindric, curving, yellow discolouring brownish with age. *Sept–Nov.*

## *T. ustale*

*Sticky cap*

**Spores** 5–6 × 3–4 μm, almost spherical

*Whitish gills spot red*

× 0.5

One of a group of *Tricholoma* spp in which the gills become spotted red. *T. ustaloides* is very similar but has a strong mealy smell, while *T. albobrunneum* (p 15) has a radially streaked cap and is usually found in coniferous woods. The dark red-brown slimy cap of *T. ustale* darkens with age. **Cap** 5–8 cm, convex then flat, smooth, with incurved margin. **Gills** sinuate, white soon spotted, crowded. **Stem** 5–7 cm, cream discolouring brownish from base up. *Sept–Oct.*

Frondose woods

## *Pleurotus dryinus*

**Spores** 9–14 × 3–5 μm

Whitish gills turn yellow with age

× 0.3
*Excentric stem*

Good to eat when young. **Cap** 9–14 cm, disc- to shell-shaped, greyish, felty, with incurved margin. **Gills** decurrent, broad, spaced. **Stem** 2–5 cm, hard, scaly, with ephemeral ring. *Sept–Dec*.

## *Panus torulosus*

**Spores** 6–7 × 3–4 μm, ovoid

Also named *P. conchatus*. On stumps; old specimens dry hard without rotting. Taste of turnip. **Cap** 5–8 cm, yellowish pink to pale brown, tough, smooth. **Gills** decurrent, pink to tan. **Stem** short, excentric, with violet down at first. *Aug–Nov*.

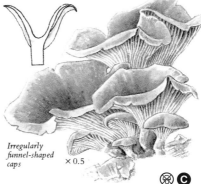

*Irregularly funnel-shaped caps* × 0.5

## *P. stipticus*

### Styptic fungus

**Spores** 3–6 × 2–3 μm, ovoid, amyloid

Common all year, forming crowded tiers on stumps. The tough flesh has a very stringent taste and is claimed to be toxic; this has not been confirmed. **Cap** 1–3 cm, roughly kidney-shaped, pale brown, thin, tough, scurfy, with incurved margin. **Gills** yellow to pale brown, thin, narrow, crowded with interveining. **Stem** reduced, compressed above. *Jan–Dec*.

*Stem broadest at point of attachment*

× 1

Frondose woods

## *Limacella guttata*

*Smooth, sticky cap*

*Spores 5–6 × 4–5 μm, almost spherical*

*Persistent ring*

Although edible with a mealy smell and taste, it is seldom found in sufficient quantities to form the basis of a meal. Generally occurs in damp woodland situations, its rather bulbous base buried in the ground. In wet weather the ring and gills may exude tear-like droplets. *Limacella* spp are closely related to the genus *Amanita*, but a volva is never found, and the sticky cap should distinguish it from a *Lepiota* sp. In older texts this fungus is known as *L. lenticularis*. **Cap** 6–10 cm, hemispherical to convex, very pale pinkish brown. **Gills** free, white, very crowded. **Stem** 8–10 cm, slightly swollen at base, colour as cap but paler, with a membranous ring which becomes tinged olive. *July–Oct.*

× 0.6

## *Chamaemyces fracidus*

*Cap surface slightly sticky*

*Spores 4–5 × 2–4 μm, ovoid, dextrinoid*

Also known as *Lepiota irrorata*. Often listed as rare but in some years it can be exceptionally abundant. Differs from closely related *Lepiota* spp mainly in structure of the pellicle. Not recommended for eating, although it is regarded as an edible fungus. **Cap** 2–5 cm, thick-fleshed, convex, cream to yellowish, with incurved margin. **Gills** free, white to cream, crowded, narrow. **Stem** 3–4 cm, white and smooth above thin, membranous ring, with yellow or brown scales below. *July–Oct.*

*Stem covered with tiny granular scales*

*White mycelial cord may be visible*

*Occurs in small groups in soil*

× 1

## *Lepiota hetieri* | *L. serena* | *L. sistrata*

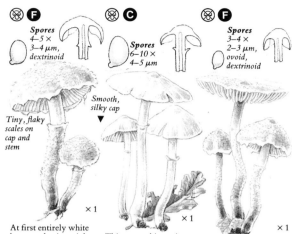

At first entirely white but soon bruises pink; finally turns brown. **Cap** 1–3 cm, convex-campanulate, creamy. **Gills** free, crowded, white. **Stem** 3–4 cm, white, pink-brown near base, scaly; ring persistent. Aug–Oct.

This pure white sp is found in small groups among dead leaves. **Cap** 2–4 cm, wavy, campanulate expanding. **Gills** free, white. **Stem** 4–6 cm, silky, with persistent thin ring. Sept–Nov.

Also called *L. seminuda*. **Cap** 1–2 cm, convex, thin, white or pinkish, scaly, with shaggy edge. **Gills** free, white. **Stem** 3–5 cm, curved, scaly, with ephemeral ring. Aug–Oct.

## *L. castanea*  *L. fulvella*

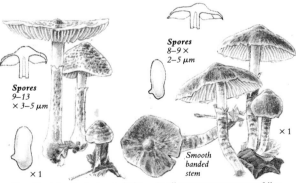

Found in rich soil, also by roadsides. Probably dangerous. **Cap** 2–4 cm, convex, umbonate, with pale yellow surface breaking up into small brown scales. **Gills** free, cream. **Stem** 3–4 cm, with swollen base, pale brown, fibrillose, with white, inconspicuous ring. Sept–Oct.

Small troops occur amongst fallen leaf litter. **Cap** 2–5 cm, umbonate, rusty to yellowish brown, with tiny, appressed scales. **Gills** free, white, crowded. **Stem** 3–6 cm, pale brown, with bands of small, flattened scales; ring inconspicuous. Sept–Oct.

## *Lepiota friesii*

*Small, pointed scales easily break off*

**Spores** 5–7 × 2–3 µm, oblong, dextrinoid

*Forking gills*

A large *Lepiota* sp, characterized by the numerous erect, pointed scales on the cap; the structure of the pellicle has caused some authorities to list it under a separate genus, *Cystolepiota*, calling it *C. aspera*. Grows in small groups on sandy soil amongst plant debris, also in gardens. The thick white flesh is bitter with a strong, unpleasant smell. **Cap** 4–10 cm, conical to convex, with crowded red-brown scales on pale brown background. **Gills** free, white, crowded, forked. **Stem** 7–10 cm, white above, brown and scaly near swollen base; ring white, rusty underneath. *Aug–Nov*.

× 0.6

## *L. bucknallii*

**Spores** 7–10 × 2–4 µm, spindle-shaped, dextrinoid

*Smells of coal gas*

× 1

A small, distinctive sp occasionally found amongst grass and dead leaves; the strong smell of coal gas recalls *Tricholoma sulphureum* (p 70), but 2 spp are not alike. Has also been placed in *Cystolepiota* (see above) on account of its pellicle. **Cap** 2–3 cm, conical to umbonate, greyish, with tiny flaky scales. **Gills** free, cream, crowded. **Stem** 2–6 cm, lavender blue, darker base. *Sept–Oct*.

## *Melanophyllum echinatum*

**Spores** 4–6 × 2–4 µm, ellipsoid, darkening to red-brown

× 1

Also known as *Lepiota haematosperma* but differs from all *Lepiota* spp in producing red-brown (not white) spores, which discolour the gills. Grows on enriched soil, especially under hedgerows; also in gardens. **Cap** 1–3 cm, brownish grey, covered with dark granular scales; margin shaggy. **Gills** free, blood red, darkening with age. **Stem** 3–6 cm, colour as cap, with granular scales. *Oct–Nov*.

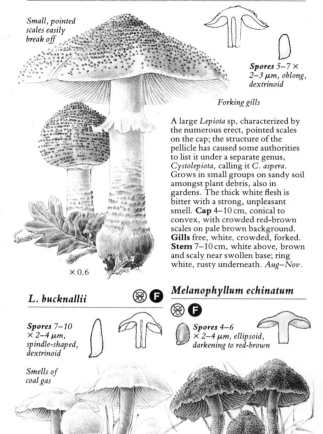

Frondose woods

## *Melanoleuca melaleuca*

*Spores 7–9 × 4–5 µm, warted, amyloid*

*Gills white, sinuate*

A very common species although rather variable, becoming much paler as it dries out. Easily confused with species of *Tricholoma*, but the *Melanoleuca* genus has quite different spores. *M. brevipes* (p 130) is similar but has a shorter stem and is less common. *M. melaleuca* also occurs on pastureland. **Cap** 4–10 cm, convex to flat, umbonate, dark brown when moist, smooth. **Gills** sinuate, white, broad, crowded. **Stem** 5–8 cm, cylindric, solid, white with longitudinal brownish fibrils. **Flesh** white to grey, soft. *Aug–Nov*.

× 0.6

## *Russula cyanoxantha*

*Thick flesh, white or greyish, with mild taste*

*Variable cap colour*

*Spores 7–9 × 6–7 µm, ellipsoid with low warts*

*Crowded, white gills*

Probably the most common and best known of all the *Russula* species, occurring throughout the north temperate zone, appearing from early July onwards. It is also one of the better edible species, with a mild, nutty taste, often sold in European markets. It is frequently consumed mixed with *Boletus* species, or else it may be pickled in vinegar. The cap colour can vary from grey, purple, olive green or brown, but the stem always remains white. **Cap** 5–15 cm, spherical then convex to depressed, with a thin margin; often greenish at the centre, sometimes uniformly green, shiny, radially veined. **Gills** adnate, white, crowded, forked. **Stem** 5–10 cm, solid, firm, white. *July–Nov*.

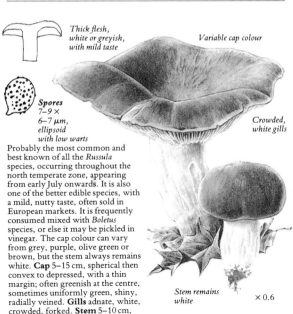

*Stem remains white*

× 0.6

Frondose woods

## *Russula heterophylla*

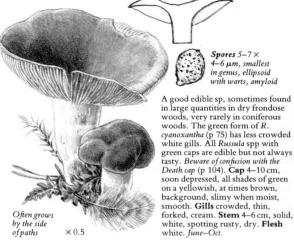

*Spores 5–7 × 4–6 μm, smallest in genus, ellipsoid with warts, amyloid*

A good edible sp, sometimes found in large quantities in dry frondose woods, very rarely in coniferous woods. The green form of *R. cyanoxantha* (p 75) has less crowded white gills. All *Russula* spp with green caps are edible but not always tasty. *Beware of confusion with the Death cap (p 104).* **Cap** 4–10 cm, soon depressed, all shades of green on a yellowish, at times brown, background, slimy when moist, smooth. **Gills** crowded, thin, forked, cream. **Stem** 4–6 cm, solid, white, spotting rusty, dry. **Flesh** white. *June–Oct.*

*Often grows by the side of paths* × 0.5

## *R. vesca*
## Bare-toothed russula

*Pellicle of cap separates at margin*

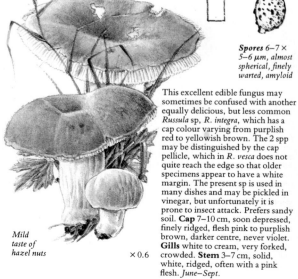

*Spores 6–7 × 5–6 μm, almost spherical, finely warted, amyloid*

This excellent edible fungus may sometimes be confused with another equally delicious, but less common *Russula* sp, *R. integra*, which has a cap colour varying from purplish red to yellowish brown. The 2 spp may be distinguished by the cap pellicle, which in *R. vesca* does not quite reach the edge so that older specimens appear to have a white margin. The present sp is used in many dishes and may be pickled in vinegar, but unfortunately it is prone to insect attack. Prefers sandy soil. **Cap** 7–10 cm, soon depressed, finely ridged, flesh pink to purplish brown, darker centre, never violet. **Gills** white to cream, very forked, crowded. **Stem** 3–7 cm, solid, white, ridged, often with a pink flesh. *June–Sept.*

*Mild taste of hazel nuts* × 0.6

Frondose woods

## *Lactarius piperatus*
### Peppery milk cap

**Spores** 6–8 × 5–6 μm, with tiny warts, amyloid

The firm white flesh produces a copious white milk with a very acrid taste which is removed by boiling. Although eaten in Russia and other eastern countries, it is picked only in dry seasons when few other spp are around, and it is not recommended for western palates. Best eaten fried but is of poor quality. Note the exceptionally crowded gills. **Cap** 7–20 cm, convex to depressed, creamy white, smooth, rigid. **Gills** white to yellow, decurrent, forked. **Stem** 3–7 cm, white, smooth. *Aug–Nov*.

× 0.5

## *L. vellereus*
### Fleecy milk cap

**Spores** 8–10 × 7–9 μm, tiny warts, amyloid

*Surface of cap soft and downy*

*Gills more spaced than L. piperatus*

The largest *Lactarius* sp with an unpalatable, hard flesh. Very conspicuous because of its size, although not always common. **Cap** 10–30 cm, soon depressed, with inrolled margin, firm, white with yellow stains. **Gills** white to cream, broadly adnate-decurrent, spaced. **Stem** 5–12 cm, stout, white or yellowish. *Aug–Nov*.

*Short, hard stem*

× 0.25

*Fruitbodies very large, producing copious acrid white milk when young*

Frondose woods

## *Lactarius subdulcis*

*Cap not zoned*

**Spores** $10–11 \times 7–9\ \mu m$, warty with fine net, amyloid

× 0.6

The copious white milk is very characteristic in having a sweetish taste that becomes slightly bitter in the mouth. This fungus is edible but not recommended—the taste has been compared to ivy. **Cap** 3–8 cm, convex then depressed, smooth, pink to pale cinnamon, darker at centre. **Gills** slightly decurrent, pinkish, crowded. **Stem** 3–5 cm, red-brown, paler above. *Aug–Oct.*

## *L. pyrogalus*

*Gills wax yellow*

**Spores** $7–8 \times 6–7\ \mu m$, incomplete net, amyloid

× 0.6

Distinguished from other greyish *Lactarius* spp by the well-spaced, wax-yellow gills. Fairly common on ground in woods, especially under hazel. The white milk is extremely acrid. **Cap** 5–10 cm, flattened to funnel-shaped, greyish brown to greyish yellow, zoned, moist but not shiny. **Gills** slightly decurrent, rather distant. **Stem** 4–7 cm, cap colour but paler. *Aug–Oct.*

## *L. flexuosus*

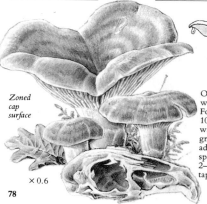

*Zoned cap surface*

**Spores** $7–8 \times 6–7\ \mu m$, with raised net

One of several grey spp with acrid white milk. Found on soil. **Cap** 5–10 cm, irregular, with wavy margin, purplish grey, velvety. **Gills** adnate, pale yellow, spaced, narrow. **Stem** 2–3 cm, short, pale, tapers below. *Aug–Oct.*

× 0.6

Frondose woods

## Hohenbuehelia petaloides

**Spores**
7–10 ×
4–5 µm,
ellipsoid

Variable in appearance with many described varieties. Usually grows in tufts either amongst woody debris, old stumps or sawdust. The spoon-shaped caps are similar in appearance to *Pleurotus* spp, but microscopically the structure is very different. The sp has a distinct smell of meal. **Cap** 2–8 cm, laterally attached and continuous with the stem, yellowish to umber, gelatinous, with a thin margin. **Gills** deeply decurrent, white to cream, thin, narrow, very crowded. **Stem** short or absent, lateral, thick, whitish, woolly. *Aug–Oct.*

× 1

*Tufted, spatula-shaped fruitbodies*

## Lentinellus cochleatus

**Spores**
4–6 × 3–5 µm,
warty, amyloid

The thin, whitish flesh tends to split very easily and has a characteristic smell of aniseed. Young specimens are edible but they soon become hard and tough. Forms large, crowded tufts on old branches and dries without rotting. **Cap** 3–7 cm, spoon-, shell- or funnel-shaped, tawny brown spotting darker, smooth. **Gills** arcuate-decurrent, whitish, narrow, very crowded. **Stem** 3–10 cm, excentric, hard, fibrous, solid, smooth, often curved, rusty brown. *Aug–Nov.*

*Smells of aniseed*

× 0.6

79

## *Entoloma aprile*

## *E. clypeatum*
### Roman-shield entoloma

**Spores** $8–12 \times 7–10\ \mu m$, polygonal

**Spores** $7–10 \times 7–9\ \mu m$, polygonal

**Flesh** whitish, smells of meal

Smooth, greasy cap

Stem with grey-brown fibrils

Robust stem, at first white

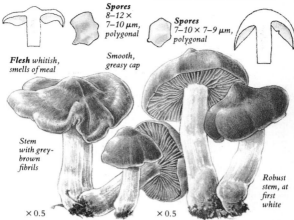

× 0.5

× 0.5

Less robust than *E. clypeatum* (right) with a greyish-brown stem. **Cap** 2–9 cm, conical to convex, umbonate, greyish brown, paler when dry. **Gills** pale to greyish pink, broad, spaced. **Stem** 5–9 cm, solid, greyish, fibrillose. *Apr–May*.

Common in troops under rosaceous shrubs. Causes stomach upset if eaten raw. **Cap** 4–10 cm, conical to convex, umbonate, greyish brown, fibrous, silky. **Gills** white to pink, broad. **Stem** 4–10 cm, solid, firm, whitish, striate. *Apr–June*.

## *E. rhodopolium*

## *E. nidorosum*

**Spores** $8–11 \times 6–9\ \mu m$

**Spores** $8–11 \times 6–9\ \mu m$, polygonal

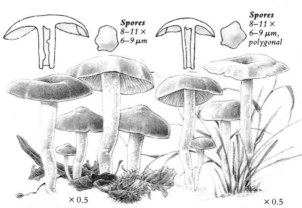

× 0.5

× 0.5

A tall, fragile sp found in troops. It has a mealy smell and is very indigestible. **Cap** 2–10 cm, silky, campanulate to depressed, yellowish grey-brown, drying paler. **Gills** adnate, white to pink, crowded. **Stem** 5–10 cm, white. *July–Sept*.

Less robust than *E. rhodopolium* (left) and with a pungent, alkaline smell. **Cap** 2–7 cm, convex to depressed, greyish brown drying to almost white, shiny, brittle. **Gills** pale to flesh pink, adnate, spaced. **Stem** 5–12 cm, slender, white. *Aug–Oct*.

Frondose woods

## *Pluteus cervinus*     *P. cinereofuscus*
### Fawn pluteus

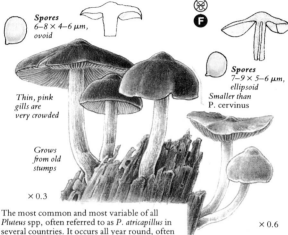

**Spores** 6–8 × 4–6 μm, ovoid

Thin, pink gills are very crowded

Grows from old stumps

× 0.3

**Spores** 7–9 × 5–6 μm, ellipsoid. Smaller than P. cervinus

× 0.6

The most common and most variable of all *Pluteus* spp, often referred to as *P. atricapillus* in several countries. It occurs all year round, often on fallen trunks. Giant specimens sometimes grow on sawdust heaps. Edible, but of poor quality. **Cap** 3–12 cm, campanulate expanding, greyish brown to umber, radially fibrillose, sticky when moist. **Gills** white to pink, free, broad. **Stem** 8–12 cm, white with dark fibrils, often swollen at base. *Jan–Dec*.

Also in leaf litter. **Cap** 3–5 cm, pale grey to grey-green, non-striate. **Gills** white then pink, crowded. **Stem** 3–8 cm, white, smooth. *Sept–Oct*.

## *P. lutescens*     *P. salicinus*
### Willow pluteus

**Spores** 5–6 × 4–5 μm, almost spherical

Gills finally pink

× 0.6

**Spores** 6–8 × 4–6 μm, ovoid

Cap and stem have greenish tints

× 0.5

Grows in troops on fallen branches. **Cap** 1–5 cm, convex, sepia brown, sometimes with yellowish margin, non-striate. **Gills** free, whitish to chrome yellow, finally pink. **Stem** 2–7 cm, cylindric, whitish but soon chrome yellow, especially at base. *May–Oct*.

Usually on rotting logs, especially willow, but occasionally on beech. **Cap** 2–6 cm, campanulate expanding, greenish to bluish grey, paler when dry, slightly wrinkled. **Gills** white then pale pink, free, crowded. **Stem** 2–7 cm, whitish, fibrillose. *Feb–Dec*.

## *Volvariella speciosa*
### Rose-gilled grisette

**Spores** 11–18 × 7–9 μm, ovoid

## *Clitopilus hobsonii*

**Spores** 6–10 × 4–6 μm, ridged

Crowded, thin, pink gills

× 0.3

A tiny, pink-spored sp, although the gills may remain white for a long time. **Cap** 0.5–2 cm, convex, kidney-shaped with lateral attachment, white to greyish cream, silky. **Gills** white to creamy pink, rather crowded. **Stem** absent or very short. **Flesh** smells and tastes of meal. *Oct–Jan*.

Resembles an *Amanita* sp, but it has pink gills and lacks a ring. Found on rich soil, often on compost heaps. Edible, but not recommended. **Cap** 8–14 cm, conical to convex, sticky, whitish grey to brownish. **Gills** free, cream then flesh pink. **Stem** 10–18 cm, satiny white, bruising brown, with persistent sac-like volva. *June–Oct*.

## *Crepidotus mollis*
### Soft slipper toadstool

**Spores** 8–9 × 5–6 μm

## *C. variabilis*

**Spores** 5–6 × 3–4 μm, warty

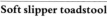

No stem

× 0.6

× 1

Very common on decaying branches. The soft, horizontal cap contains a gelatinous layer giving an elastic texture when pulled apart. **Cap** 2–7 cm, convex, kidney-shaped, laterally attached, pale ochre brown with scattered brown fibrillose scales. **Gills** cinnamon brown, crowded. **Stem** absent. *May–Dec*.

This small fungus may be found all year round in troops on dead twigs and straw, although it is more common in the autumn. Many of these small white spp are only reliably distinguished by their spore characteristics. **Cap** 1–2 cm, kidney-shaped, laterally attached, white, felty. **Gills** white, dull yellow to brown, rather spaced. *Aug–Oct*.

Frondose woods

## *Cortinarius ochroleucus*

**Spores** 7–8 × 4–5 µm, warty

A large, inedible sp with a bitter taste; sometimes occurs in clusters. Becomes sticky in wet weather. **Cap** 3–8 cm, umbonate expanding, pallid with ochre-yellow centre, silky, soon dry. **Gills** adnate, clay to ochre yellow, crowded. **Stem** 2–9 cm, whitish, with cortinoid zone near apex. *Aug–Oct.*

*Tapering stem base* × 0.5

## *C. trivialis*

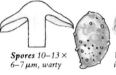

**Spores** 10–13 × 6–7 µm, warty

*Veil breaks into bands*

Similar to *C. collinitus* (p 29) but with duller colours. Recognized by the slimy bands on the stem, resulting from the breakdown of the veil. **Cap** 3–11 cm, convex, expanding, date brown to bay brown, very slimy. **Gills** clay to dull rusty brown, crowded. **Stem** 5–12 cm, tapers at both ends, with whitish scales on a darker background. *Aug–Oct.*

× 0.5

## *C. largus*

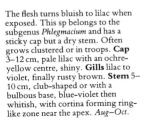

**Spores** 10–12 × 5–7 µm, almond-shaped

The flesh turns bluish to lilac when exposed. This sp belongs to the subgenus *Phlegmacium* and has a sticky cap but a dry stem. Often grows clustered or in troops. **Cap** 3–12 cm, pale lilac with an ochre-yellow centre, shiny. **Gills** lilac to violet, finally rusty brown. **Stem** 5–10 cm, club-shaped or with a bulbous base, blue-violet then whitish, with cortina forming ring-like zone near the apex. *Aug–Oct.*

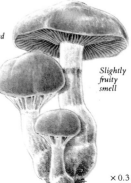

*Slightly fruity smell*

× 0.3

### *Inocybe geophylla*
#### Common white inocybe

**Spores** 8–10 × 4–5 μm, ellipsoid

### *I. geophylla* var *lilacina*
#### Lilac inocybe

**Spores** 8–10 × 4–5 μm

### *I. griseolilacina*

**Spores** 8–12 × 4–6 μm

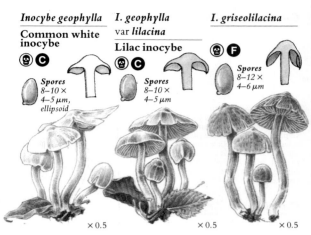

A silky white sp, with an earthy smell. **Cap** 1–3 cm, conical with incurved margin, white, fibrillose. **Gills** adnexed, ochre yellow to clay brown, crowded. **Stem** 3–5 cm, white, silky, with small bulbous base. *June–Nov*.

This variety of *I. geophylla* is identical to the typical form (left) except for its uniform lilac colour. Both grow in troops and are often found together. The clay-brown gills separate it from the Amethyst deceiver (p 66). *June–Nov*.

Found amongst leaves; also by roadsides. Has a mealy smell. **Cap** 1–3 cm, umbonate, at first lavender blue then brown, shaggy-scaly. **Gills** adnate, whitish soon brown. **Stem** 4–6 cm, azure blue, with white fibrils. *Aug–Nov*.

### *I. eutheles*

**Spores** 7–11 × 4–6 μm, ellipsoid

### *I. pyriodora*
#### Pear-scented inocybe

**Spores** 9–13 × 5–7 μm, almond-shaped

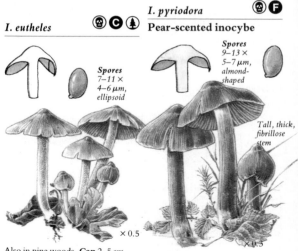

*Tall, thick, fibrillose stem*

Also in pine woods. **Cap** 2–5 cm, umbonate, finely fibrillose, pale brown, with paler margin. **Gills** white to clay brown. **Stem** 5–7 cm, whitish, powdery. *Aug–Nov*.

Smell of ripe pears. **Cap** 3–7 cm, umbonate, ochre brown, fibrillose. **Gills** yellow-brown, crowded. **Stem** 6–10 cm, reddish. *Aug–Nov*.

## *I. cookei*  ## *I. flocculosa*  ## *I. godeyi*

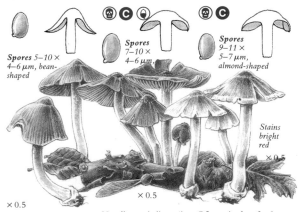

**Spores** 5–10 × 4–6 μm, bean-shaped

**Spores** 7–10 × 4–6 μm

**Spores** 9–11 × 5–7 μm, almond-shaped

Stains bright red

Reportedly not toxic, but best avoided. **Cap** 3–5 cm, conical to campanulate, pale ochre, fibrillose. **Gills** whitish then cinnamon, crowded. **Stem** 3–6 cm, with swollen base, whitish to cream. *Aug–Nov.*

Usually on chalky soil, in small troops. **Cap** 2–4 cm, brown with greyish fibrils, campanulate. **Gills** crowded, ochre yellow. **Stem** 3–5 cm, cylindric, whitish brown; powdery apex. *Sept–Nov.*

Often mistaken for *I patouillardii* (p 135). Smell earthy or mealy. **Cap** 3–5 cm, conical, silky, cream discolouring red. **Gills** adnexed, white to cinnamon. **Stem** 4–6 cm, white, reddening, with a bulbous base. *Aug–Nov.*

## *I. asterospora*   ## *I. lanuginosa*

Fibrillose cap

**Spores** 9–13 × 6–10 μm, spiny

**Spores** 7–10 × 4–7 μm, with nodules

Often found on wet clay soils. **Cap** 3–5 cm, umbonate, yellowish brown with darker fibrils. **Gills** grey then cinnamon, crowded. **Stem** 4–7 cm, red-brown, with white bulbous base. *Sept–Nov.*

Distinguished by the conspicuous spiky scales on the cap. Prefers damp woods. **Cap** 1–2 cm, convex to flat, brown, fibrillose. **Gills** adnexed, cream to brown. **Stem** 3–4 cm, brown, scaly. *Aug–Nov.*

## Inocybe calospora

**Spores**
$8-11 \times 7-9\,\mu m$, spherical, with long spines

## Galerina mniophila

**Spores**
$9-13 \times 5-8\,\mu m$, ovoid, faintly roughened

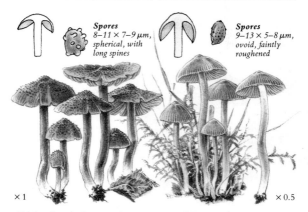

Chiefly of interest because of its unusual spores. Solitary or in small groups, amongst dead leaves. **Cap** 1–2 cm, umbonate, chestnut brown, with minute erect scales. **Gills** adnexed. **Stem** 3–7 cm, dark brown, with white base. *Aug–Nov.*

A small, honey-coloured sp, often amongst moss (not *Sphagnum*). **Cap** 1–3 cm, campanulate, yellow-brown, striate. **Gills** adnate, ascending, ochre to brownish black, fairly spaced. **Stem** 2–7 cm, yellowish with white base, striate. *Sept–Nov.*

## Hebeloma sinapizans

*Section always shows V-shaped flesh at top of stem cavity*

**Spores**
$11-14 \times 6-8\,\mu m$, almond-shaped, roughened

*Finely toothed gill edge*

*Slightly sticky cap*

The white, finely toothed gill edge distinguishes a *Hebeloma* sp from most *Cortinarius* spp. All *Hebeloma* spp cause gastro-intestinal upset and should not be eaten. This is the largest of the genus, with a distinct smell of raw potatoes. The upper stem breaks up into small curved scales (not granular as in *H. crustuliniforme*). **Cap** 7–15 cm, fleshy, convex, pale pinky brown, smooth. **Gills** sinuate, cinnamon, crowded. **Stem** 7–12 cm, white, cylindric, scaly. *Sept–Oct.*

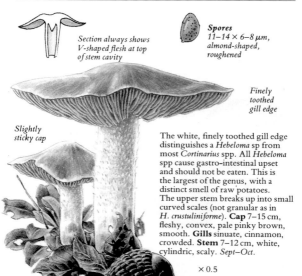

Frondose woods

## *H. crustuliniforme*
### Fairy cake hebeloma

**Spores**
$10–14 \times 5–7\,\mu m$, almond-shaped, roughened

Also known as Poison pie, for it has a bitter taste, and may cause vomiting, cramp and diarrhoea. It is the most common *Hebeloma* sp, but varies in size and colour. Note the water drops on the gill edge and smell of radish. **Cap** 5–10 cm, fleshy, convex, cream to pale olive brown, smooth, sticky. **Gills** clay to cinnamon, crowded. **Stem** 4–8 cm, cylindric, whitish, with granules near apex. *Aug–Nov*.

× 0.5

## *H. longicaudum*

**Spores**
$11–13 \times 5–7\,\mu m$, almond-shaped, roughened

Similar to *H. crustuliniforme* (above) in many ways, but this sp is more slender and the stem length always exceeds the cap diameter. Grows in damp places. **Cap** 4–8 cm, convex, umbonate, uniformly yellowish brown, sticky, smooth. **Gills** sinuate, whitish to cinnamon. **Stem** 8–11 cm, cylindric, white becoming tawny brown from the base up. *Sept–Nov*.

*Long, slender stem*

*Lacks a distinctive smell*

× 0.5

## *H. sacchariolens*
### Scented hebeloma

**Spores**
$13–17 \times 7–10\,\mu m$, almond-shaped

Distinguished from all other spp by a strong, sickly smell, which has been variously described as recalling burnt sugar, orange flowers, cheap scented soap and even "harlots"! It grows in small groups, preferring damp sandy soil. **Cap** 2–5 cm, convex, whitish then brown with age, slightly sticky, smooth. **Gills** sinuate, cinnamon brown. **Stem** 4–5 cm, white, tawny brown at base, striate. *Sept–Oct*.

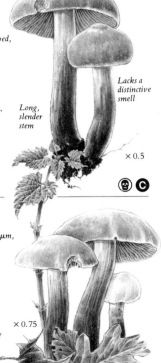

× 0.75

87

## *Gymnopilus junonius*
## Orange pholiota

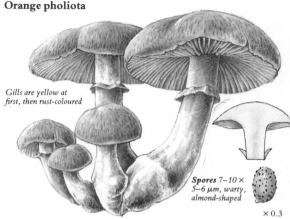

*Gills are yellow at first, then rust-coloured*

*Spores 7–10 × 5–6 µm, warty, almond-shaped*

× 0.3

A handsome, golden-yellow fungus that is found clustered on stumps and at tree bases, especially ash and apple. The thick, yellow flesh is tough and bitter-tasting. Previously known as *Pholiota spectabilis*. **Cap** 6–13 cm, tawny yellow to golden brown, convex, fleshy, with an incurved margin, dry, radially fibrillose. **Gills** crowded, broadly adnate. **Stem** 6–14 cm, swollen, tapers to a short "root", fibrillose, with a high, spreading ring. *Aug–Dec*.

## *Phaeolepiota aurea*

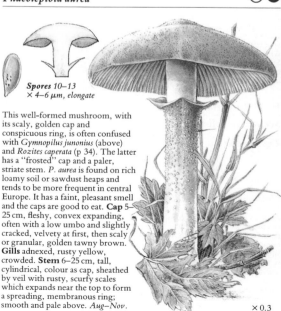

*Spores 10–13 × 4–6 µm, elongate*

This well-formed mushroom, with its scaly, golden cap and conspicuous ring, is often confused with *Gymnopilus junonius* (above) and *Rozites caperata* (p 34). The latter has a "frosted" cap and a paler, striate stem. *P. aurea* is found on rich loamy soil or sawdust heaps and tends to be more frequent in central Europe. It has a faint, pleasant smell and the caps are good to eat. **Cap** 5–25 cm, fleshy, convex expanding, often with a low umbo and slightly cracked, velvety at first, then scaly or granular, golden tawny brown. **Gills** adnexed, rusty yellow, crowded. **Stem** 6–25 cm, tall, cylindrical, colour as cap, sheathed by veil with rusty, scurfy scales which expands near the top to form a spreading, membranous ring; smooth and pale above. *Aug–Nov*.

× 0.3

Frondose woods

## *Agrocybe cylindracea*

**Spores**
$8–10 \times 5–6\ \mu m$, with germ-pore

Found tufted on old roots, especially poplar and willow. Smells of old wine, but is good to eat. Also known as *A. aegerita*. **Cap** 4–12 cm, rounded then convex, with a dry, silky-smooth surface which is often cracked; clay brown initially, paling to light tan, almost white at margin. **Gills** decurrent to adnate, cream to cinnamon brown, thin. **Stem** 5–12 cm, cylindric, fibrillose, white, with a thick, pendent ring which turns from white to brown. *May–Nov.*

× 0.3

## *Kuehneromyces mutabilis*
### Two-toned pholiota

**Spores**
$6–8 \times 3–5\ \mu m$, ovoid, truncated by a germ-pore

Recognized by the two-toned cap that absorbs water easily and changes colour as it dries. Grows in dense clusters on stumps and may be confused with the Honey fungus (p 45). The caps are edible, but reject fibrous stems. **Cap** 4–8 cm, date brown when wet, drying yellow from centre, margin striate. **Gills** clay to rust brown, adnexed, thin, crowded. **Stem** 5–8 cm, rust brown, curved, scaly up to membranous ring. *Apr–Dec.*

× 0.5

## *Pholiota gummosa*

**Spores**
$5–7 \times 3–4\ \mu m$, ellipsoid, with germ-pore

In small clusters on old stumps. Distinguished from *Hypholoma fasciculare* (p 58) by its sticky, green-tinted cap and the lack of green in the gills. **Cap** 3–6 cm, campanulate becoming flat, pale yellow with green tints, fleshy, glutinous, with small, white scales. **Gills** adnate, yellow to brown, narrow, crowded. **Stem** 4–7 cm, pale yellow with rust-brown base, fibrillose. *Oct–Dec.*

× 0.5

Frondose woods

## *Pholiota squarrosa*
# Shaggy pholiota

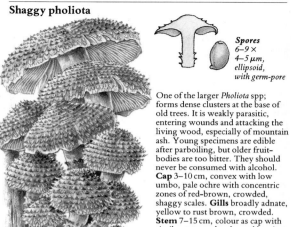

**Spores**
$6–9 \times 4–5\,\mu m$, ellipsoid, with germ-pore

One of the larger *Pholiota* spp; forms dense clusters at the base of old trees. It is weakly parasitic, entering wounds and attacking the living wood, especially of mountain ash. Young specimens are edible after parboiling, but older fruit-bodies are too bitter. They should never be consumed with alcohol. **Cap** 3–10 cm, convex with low umbo, pale ochre with concentric zones of red-brown, crowded, shaggy scales. **Gills** broadly adnate, yellow to rust brown, crowded. **Stem** 7–15 cm, colour as cap with similar upturned scales and bearing a small, dark, fibrous ring on the upper part. *July–Dec.*

× 0.3

## *P. aurivella*

*Slimy cap*

**Spores**
$7–9 \times 5–6\,\mu m$, ellipsoid, with germ-pore

*Cap covered in large, flat, gelatinous, red-brown scales*

*Dry stem, with small, fibrillose scales*

A large species, easily recognized, usually forming small clusters on the upper branches of dead or living broad-leafed trees. The slimy cap distinguishes it from *P. squarrosa* (above) and the dry stem from *P. adiposa* (p 121). Although edible, it has a bitter, unpleasant taste. **Cap** 7–12 cm, bright yellow, very slimy, with zones of large, flattened, chestnut-brown scales, which are more abundant at the centre. **Gills** pale yellow then rust brown, adnate, crowded. **Stem** 10–15 cm, yellowish, rust brown towards base, dry, covered with zones of small, recurved, sticky scales; ring small, fibrillose, soon disappearing. *Sept–Nov.*

*Grows on upper tree branches*

× 0.3

Frondose woods

## *Agaricus placomyces*

**Spores**
$4–6 \times 3–4\,\mu m$, ovoid

Sooty-brown scales on cap surface

All parts bruise yellow

Also known as *A. meleagris*, it is closely related to the toxic *A. xanthodermus* (p 136) and should be avoided. It is recognized by the flattened cap, which bruises deep yellow, and the unpleasant carbolic smell. The stem base turns bright yellow when broken. *A. silvaticus* (p 36) is also similar but has a deeply reddening flesh. **Cap** 5–12 cm, campanulate then flat, with a dark, sooty-brown centre and numerous small scales on a white background. **Gills** free, bright pink then blackish brown, crowded. **Stem** 6–10 cm, tall, cylindric with bulbous base, white bruising bright yellow; ring large. *July–Sept*.

× 0.5

## *Stropharia aeruginosa*
### Verdigris toadstool

**Spores**
$7–9 \times 4–5\,\mu m$, ovoid

## *Coprinus micaceus*
### Glistening ink cap

**Spores**
$7–10 \times 5–7\,\mu m$, with germ-pore

Slimy cap

× 1

× 0.5

Exceptionally attractive when young, but the vivid colours soon fade. Common in grassy woods. **Cap** 2–7 cm, convex, umbonate, blue-green, with small white scales. **Gills** adnate, grey-brown. **Stem** 4–7 cm, slimy, paler than cap, with white, flaky scales; ring black, white below. *July–Nov*.

Arises from buried wood or dead roots in dense tufts. Common everywhere. **Cap** 2–4 cm, fragile, ovoid to campanulate, ochre to tawny brown, grooved almost to centre, with glistening surface. **Gills** crowded, pale to dark brown, deliquescent. **Stem** 5–8 cm, white, smooth. *Jan–Dec*.

Frondose woods

## *Coprinus disseminatus*

### Trooping crumble cap

*Always in large numbers*

**Spores**
$9–11 \times 5–7\,\mu m$,
ellipsoid,
with germ-pore

Common from spring onwards; hundreds of individuals can appear overnight, densely clustered on rotten wood. The gills do not deliquesce like other ink caps. **Cap** 0.5–1.5 cm, campanulate, thin, fragile, pale yellow then grey with yellow centre, grooved, glistening. **Gills** adnate, white to grey, narrow, crowded. **Stem** 1–4 cm, curved, white. *Apr–Nov.*

×0.5

## *C. lagopus*

**Spores**
$10–11 \times 5–8\,\mu m$, with germ-pore

*Gills soon deliquescent*

## *Psathyrella candolleana*

### Fringed crumble cap

**Spores**
$7–9 \times 4–5\,\mu m$, with germ-pore

×0.5

Usually solitary on soil amongst leaves in shady places. A fragile sp with a hollow stem that rapidly elongates. **Cap** 2–3 cm, ovoid at first, then flat with upturned margin, grey, radially grooved, with small white fibrils. **Gills** very narrow, soon black, crowded. **Stem** 10–15 cm, thicker near base, white, covered with fine, white fibrils. *July–Oct.*

May be seen in large quantities in early summer, often on or around stumps in moist, shady places. The young gills are a distinctive pale pinkish grey and the cap is smooth and rather brittle. **Cap** 2–5 cm, campanulate expanding, pale yellowish brown drying to almost white, glistening; margin shaggy. **Gills** adnexed, finally dark brown. **Stem** 5–8 cm, white, fragile, fibrillose. *Apr–Nov.*

## P. hydrophila

**Spores** 4–7 × 3–4 µm, with germ-pore

Forms dense tufts on stumps and old roots. The cap is much paler when dry, and is easily confused with *Kuehneromyces mutabilis* (p 89), but the white, non-scaly stems are diagnostic. Edible but poor. **Cap** 2–5 cm, campanulate, not expanding, date brown to pale tan, smooth, shiny, cracking when old. **Gills** adnate, grey to date brown, crowded. **Stem** 3–8 cm, rigid, white, striate. *Aug–Dec.*

× 0.5

*Smooth white stem*

## P. spadiceogrisea

**Spores** 7–10 × 4–5 µm, with germ-pore

Appears in spring, forming troops (but not tufts) around old stumps and tree bases. Edible but not recommended. **Cap** 2–6 cm, campanulate expanding to flat, date brown drying whitish, fragile, with a striate margin. **Gills** adnexed, blackish brown, narrow, crowded. **Stem** 4–8 cm, cylindric, white, shiny. *Mar–Nov.*

× 0.3

## P. conopilea

**Spores** 12–16 × 7–9 µm, ellipsoid, with germ-pore

Sometimes found on grassy banks, but more typically found in clusters around the base of stumps and trees, often reappearing every year. The dark brown cap is very brittle and often cracks. Also known as *P. subatrata*. **Cap** 2–5 cm, conical or campanulate, not expanding, bay brown drying pale ochre yellow, fragile. **Gills** adnexed, flesh pink to blackish brown. **Stem** 10–15 cm, long, cylindric, silvery, shiny, smooth. *Sept–Nov.*

*Conical cap*

× 0.5

Frondose woods

## Psathyrella spadicea

**Spores**
$7-10 \times 4-5 \, \mu m$,
with germ-pore

*Often in clusters on old roots*

*Smooth stem*

× 0.3

Most *Psathyrella* spp have dark caps which rapidly dry out and become very pale. This sp is difficult to identify because its gills finally turn reddish brown, rather than purplish black. Its main characters are gill colour, the inrolled cap margin when young, and the tufted habit on wood (although it also occurs singly on the ground). It differs from *P. hydrophila* (p 93) in colour of gills and absence of veil in young specimens. **Cap** 3–10 cm, convex expanding, date brown drying paler, smooth, finally cracking. **Gills** adnexed, pinkish brown to umber. **Stem** 4–10 cm, cylindric, white, silky. *Apr–Nov*.

## Strobilomyces floccopus
### Old man of the woods

**Spores**
$9-15 \times 8-12 \, \mu m$,
with network ornament

*Thick, woolly scales on cap*

*Pale grey pores bruise reddish*

*Elongate, scaly stem*

*Usually solitary*

× 0.5

The name *Strobilomyces* means "pine cone fungus", referring to the pyramidal shape of the greyish-black, overlapping scales which cover the cap and stem. Young specimens are entirely white and can be eaten, but they soon blacken and become too tough for the table. This sp is one of the easiest boletes to recognize. **Cap** 4–10 cm, hemispherical to convex with shaggy incurved margin, dry, covered with large, blackish-brown scales. **Tubes** adnate-decurrent, greyish white, redden on exposure. **Pores** large, angular, greyish. **Stem** 8–14 cm, cylindric, hard, colour as cap. **Flesh** soft, moist, white, discolours red, finally black. *Aug–Oct*.

# Birch woods

As the birch is such an attractive tree, it is frequently planted in urban areas and gardens. The small leaves cast a light shade and the leaf litter breaks down into a fertile humus so there is often a rich undergrowth. In nature the birch is quick to invade bare ground or cleared areas, especially where the soil is poor. Few birches, however, grow to a large size and they are often attacked by the heartrot bracket fungus *Piptoporus betulinus*. Among the numerous mushrooms and toadstools associated with birch are many species of *Amanita*, *Lactarius*, *Cortinarius* and *Russula*, including the very common Fly agaric (below) and Grisette (p 96).

## *Amanita muscaria*

### Fly agaric

*Spores* 
$10–12 \times 7–9 \, \mu m$, 
non-amyloid

White, free gills

White ring near stem apex

The best known toadstool, often illustrated in children's books. Only lethal if eaten in enormous amounts, but presence of muscimol causes psychotropic poisoning similar to alcoholic intoxication, and can lead to coma in extreme cases. Subject of much ancient mythology, and said to be "Soma", the divine mushroom of immortality of early Eurasian religions. Used as an intoxicant in northern Europe and Siberia before the introduction of vodka. **Cap** 15–25 cm, scarlet to pale orange, with large white loose scales of the veil. **Gills** crowded. **Stem** 12–25 cm, cylindric with bulbous base, white; ring large, pendent. *Aug–Nov*.

*Volva reduced to zones of white scales*

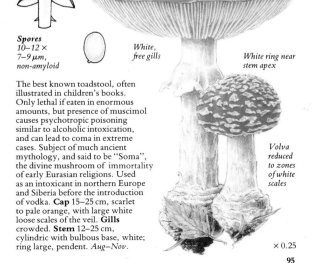

× 0.25

## Amanita fulva
### Tawny grisette

*Cap surface smooth, shiny*

**Spores** 9–13 μm diam, spherical, non-amyloid

*No ring*

*Striate margin*

*Arises from cup-like volva*

This graceful, slender fungus is common and widespread in woods, and found from summer onwards. More frequent and usually smaller than the Grisette (below), it is good to eat, but because of its thin flesh it is often used in mixed mushroom dishes. It contains haemolysins and therefore must be cooked first. Do not confuse with other *Amanita* spp from which the ring has been rubbed off. Grows amongst leaf mould, often by the side of paths. **Cap** 4–10 cm, campanulate to depressed with a slight umbo, tawny brown. **Gills** free, white to yellowish. **Stem** 10–20 cm, long, paler than cap, slightly scaly, arising from a pale tawny, membranous volva. *May–Nov*.

× 0.3

## A. vaginata
### Grisette

*Cap surface free of scales*

**Spores** 10–13 μm diam, spherical, non-amyloid

*Ring absent*

*Long, slender stem*

A widespread species throughout the northern hemisphere, even extending into the Arctic Circle, although it may be locally rare. Sometimes found in coniferous woods. The grisettes were formerly placed in the genus *Amanitopsis* owing to the absence of a ring on the stem, but it is now known that a ring-like structure is present in the early stages of development. Many varieties have been described for this species based on differences in cap colour, which may be whitish to yellowish grey. The grey form shown here is said to be good to eat after cooking, but as with the Tawny grisette, its thin flesh is very fragile and difficult to carry. **Cap** 3–8 cm, campanulate to depressed, often umbonate, smooth; grooved margin. **Gills** white, free. **Stem** 10–20 cm, hollow, whitish with pale grey zones; volva white, cup-like, tall. *June–Oct*.

× 0.3

Birch woods

## *Tricholoma album*  ## *T. fulvum*

 **Spores** 4–7 × 3–5 µm, ovoid

 **Spores** 5–7 × 4–6 µm, ellipsoid

*Nauseating smell*

*Variable in shape*

*Unpleasant smell of rancid meal*

*Pale yellow gills are spotted red* ▼

× 0.5

× 0.5

A white, fleshy sp, sometimes robust, sometimes slender. Has an unpleasant smell and bitter taste. **Cap** 5–12 cm, convex expanding, pure white with yellowish centre, smooth, dry. **Gills** sinuate, white, crowded. **Stem** 6–8 cm, white, slightly fibrous, with swollen base. **Flesh** white, spongy. *Sept–Nov.*

Probably the commonest of the brown *Tricholoma* spp, it is distinguished by its slimy cap and yellowish gills. Also known as *T. flavobrunneum*. **Cap** 7–12 cm, convex to slightly umbonate, reddish brown, radially fibrillose. **Gills** sinuate, crowded. **Stem** 7–12 cm, tapers above, slimy at first, reddish brown, scaly below. **Flesh** white, yellow in stem. *Sept–Nov.*

## *Lactarius turpis*

## Ugly milk cap

 **Spores** 6–8 × 5–6 µm, amyloid

A large mushroom with a short, squat stem, often in large numbers in damp soil, but always with birch. May be overlooked because of its sombre colours. Also known as *L. necator* and *L. plumbeus*. **Cap** 6–18 cm, convex to depressed, olive brown to blackish, slimy, with an inrolled margin. **Gills** cream to straw yellow, spotted brown on bruising, crowded. **Stem** 4–6 cm, colour as cap, soon hollow, sticky. **Flesh** whitish, with copious white milk, acrid taste. *July–Oct.*

× 0.3

97

Birch woods

## *Lactarius tabidus*

 **Spores** $7-9 \times 5-7\,\mu m$, warty with network, amyloid

Acrid white milk turns yellow

×0.5

Easily mistaken for *L. subdulcis* (p 78) but has yellowing milk. Often in *Sphagnum* moss. **Cap** 2–5 cm, rusty brown, umbonate with small central nipple, smooth with striate margin. **Gills** decurrent, pale brown, crowded. **Stem** 2–4 cm, pale above. Aug–Oct.

## *L. vietus*
### Grey milk cap

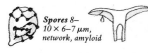 **Spores** $8-10 \times 6-7\,\mu m$, network, amyloid

White milk dries grey

In large numbers on damp ground

×0.5

Edible only after boiling because of acrid taste. **Cap** 5–8 cm, soon depressed, pale purplish grey to pinkish yellow. **Gills** adnate, white to yellow, thin, crowded. **Stem** 5–7 cm, pale. Sept–Oct.

## *L. uvidus*

 **Spores** $8-10 \times 7-8\,\mu m$, raised network, amyloid

Bruises lilac

×0.5

In small groups in damp places, always under birch. The white flesh produces an acrid white milk which turns lilac on exposure. **Cap** 4–10 cm, convex, flat or depressed, pale grey tinted lilac, sticky. **Gills** adnate to decurrent, white, crowded. **Stem** 6–9 cm, pale. Aug–Nov.

## *L. glyciosmus*
### Coconut-scented milk cap

**Spores** $8-9 \times 5-6\,\mu m$, ovoid, amyloid

Strong smell of coconut oil

×0.5

Similar to *L. vietus* (above) but the milk remains white. **Cap** 2–6 cm, convex to depressed with central nipple, greyish lilac, slightly scaly. **Gills** decurrent, yellowish pink to greyish, narrow. **Stem** 3–7 cm, white to pale yellow, bruises brown, spongy. Aug–Nov.

Birch woods

## *Russula claroflava*
### Yellow swamp russula

## *R. versicolor*

**Spores** 9–10 × 7–8 μm, warty, amyloid

Gills whitish, crowded

× 0.3

Usually found amongst moss. The slimy, bright yellow cap fades with age. Mild-tasting, with a strong, fruity smell. **Cap** 5–10 cm, convex to depressed, shiny. **Stem** 5–9 cm, white to greyish. *Aug–Sept.*

**Spores** 6–10 × 4–7 μm, partial network, amyloid

Acrid taste, esp in gills

× 0.5

Although the cap colour is variable (hence the name), there is always a greyish-green tinge. A slender, fragile sp. **Cap** 2–5 cm, convex to slightly depressed, thin, slimy at first, rose, violet or brownish. **Gills** adnate, cream, thin, crowded. **Stem** 2–5 cm, whitish. *July–Oct.*

## *R. nitida*

## *R. betularum*

**Spores** 8–10 × 7–8 μm, amyloid

Slender, white stem

× 0.5

The white flesh is inedible because of its acrid taste. Often mistaken for *R. fragilis* (p 53), but it has a relatively longer stem and generally more yellowing in the cap. Fairly common in grassy copses. **Cap** 2–5 cm, convex, rose pink, often with yellowish centre; margin grooved. **Gills** adnexed, white, spaced. **Stem** 3–6 cm, white. *Aug–Oct.*

**Spores** 8–10 × 6–9 μm, warty, amyloid

Gills ochre yellow ▶

× 0.5

Often in *Sphagnum* moss. A small to medium sp, not eaten because the flesh is too thin. **Cap** 3–7 cm, finally deeply depressed, shiny, wine red or purplish, sometimes with ochre or olive tints, darker at centre; margin grooved. **Gills** pale cream to ochre, fairly spaced. **Stem** 4–8 cm, white flushed with pale pink. *July–Oct.*

Birch woods

## *Cortinarius pholideus*

*Tiny scales point upwards*

**Spores** 6–9 × 5–6 μm, warty

*Gills and stem apex at first blue-violet*

Although the taste is mild there is little to recommend this fungus for the table. It is best to avoid the many *Cortinarius* spp as some are extremely poisonous and even fatal. This pretty, relatively slender sp with tawny scales on the cap and stem is placed in the subgenus *Sericeocybe* because it does not change colour in wet weather. Common under birch on heathy soil. **Cap** 3–6 cm, convex to depressed, fawn to pale date brown, with small, fibrillose, dark brown scales. **Gills** pale lilac soon rusty, crowded. **Stem** 6–12 cm, pale buff, scaly. **Flesh** blue-violet, fading whitish. *Aug–Sept*.

× 0.5

## *C. triumphans*

*Slimy, yellowish cap*

**Spores** 12–17 × 6–7 μm, elongate spindle-shaped, warty

A large species with a slimy cap and dry stem, features which are typical of the subgenus *Phlegmacium*. It is recognized by the ochre-yellow, ring-like bands on the stem and the light yellow veil that initially covers the gills. *C. crocolitus*, also of birch woods and probably more common in Britain, has a yellower, smoother cap and the gills have a blue tinge at first. A useful field test for *C. triumphans* is to apply sodium hydroxide solution to the flesh, which discolours a bright golden yellow. **Cap** 3–12 cm, yellow-brown to tawny ochre, almost orange at centre, with small, appressed scales. **Gills** clay to ochre brown. **Stem** 8–12 cm, thick, swollen at base, at first with a white veil that breaks down to form ring-like yellowish zones on a paler background. *Aug–Oct*.

× 0.3

Birch woods

## *C. hemitrichus*
### Scurfy cortinarius

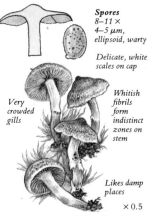

*Spores* 8–11 × 4–5 µm, ellipsoid, warty

Delicate, white scales on cap

Very crowded gills

Whitish fibrils form indistinct zones on stem

Likes damp places

× 0.5

Always with birch, often forming large troops on heathy soil. *C. paleaceus* (p 155) is similar but has a smell of geraniums. The present sp is identified by the coating of white, powdery scales over the cap surface. **Cap** 2–5 cm, convex with small umbo, deep brown, covered with whitish scales that soon fall off. **Gills** brown, crowded. **Stem** 5–7 cm, slender, fibrillose, greyish. *Apr–Nov.*

## *C. armillatus*
### Red-banded cortinarius

*Spores* 9–12 × 5–6 µm

Stem with 1 or more red zones

Broad, widely spaced gills

× 0.3

On acid soil, often near swamps. Avoid spp with red-banded stems as some, such as *C. speciosissimus* (p 30), are deadly. **Cap** 5–10 cm, convex, fleshy, brick red to tan, darker centre. **Gills** pale cinnamon to bay brown. **Stem** 6–15 cm, red-brown, fibrillose. *Aug–Oct.*

## *Paxillus involutus*
### Brown roll-rim

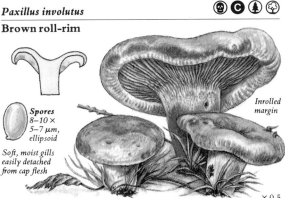

*Spores* 8–10 × 5–7 µm, ellipsoid

Soft, moist gills easily detached from cap flesh

Inrolled margin

× 0.5

An extremely common sp, especially associated with birch on poor soils. Stated to be harmless when eaten occasionally, but there is a cumulative poisonous effect if eaten repeatedly over a period of years which has proved fatal. **Cap** 5–25 cm, convex soon depressed, felty, sticky when wet, olive brown to rusty brown. **Gills** decurrent, cream to red-yellow, rust-spotted, crowded. **Stem** 3–7 cm, firm, pale, bruising dark brown. *June–Dec.*

Birch woods

## *Leccinum scabrum*
### Brown birch boletus

*Rough scaly stem*

× 0.5

**Spores**
$13–18 \times 5–6\ \mu m$, spindle-shaped

*Leccinum* spp have scaly stems and are often called "Rough stalks". All are excellent to eat, though often maggoty, and are important economically in Nordic countries as a food source for reindeer. *L. holopus* is a whitish sp of birch swamps, while *L. vulpinum* has a purplish-brown cap and grows under pine. *L. scabrum* differs from other birch spp by its brown cap, blackish stem scales and white, unchanging flesh. **Cap** 4–10 cm, uniformly yellowish brown or with grey tints, velvety then smooth. **Tubes** whitish, depressed around stem. **Pores** greyish white. **Stem** 10–20 cm, firm, whitish with dark granular scales. *June–Oct.*

## *L. versipelle*
### Orange birch boletus

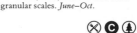

*White flesh blackens on exposure*

**Spores**
$12–19 \times 3–5\ \mu m$, spindle-shaped

*Tall stem with blackish scales on a white background*

× 0.5

Also known as *L. testaceoscabrum*, it is recognized by its orange cap and scaly stem. Often confused with *L. aurantiacum* which grows under poplar and has orange-brown stem scales. Although edible, the flesh discolours to a dark purplish grey. **Cap** 8–20 cm, orange to yellowish red or apricot, hemispherical expanding to convex, dry. **Pores** greyish, minute, round. **Stem** 10–20 cm, often club-shaped, striate, with dark scales. *June–Oct.*

# Beech and oak woods

Beech and oak woods provide the oldest and most natural forest in central western Europe, and it is here that the greatest variety of fungi will be found. Oaks are sturdy, long-lived trees and thrive in deep, heavy soils, whereas the tall, smooth-barked beech competes more successfully on shallow, chalky soils. The most important toadstool one should recognize in these woods is the deadly Death cap (p 104), which is replaced further north by the Destroying angel (p 104). Excellent edible species include Chanterelle (below), Oyster mushroom (p 110) and Cep (p 123).

## *Cantharellus cibarius*

### Chanterelle

*Spores 7–11 × 4–7 μm*

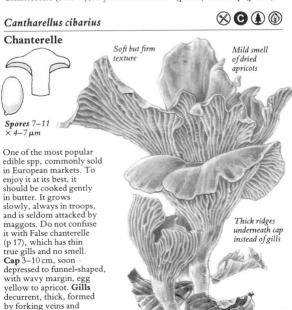

*Soft but firm texture*

*Mild smell of dried apricots*

*Thick ridges underneath cap instead of gills*

×1

One of the most popular edible spp, commonly sold in European markets. To enjoy it at its best, it should be cooked gently in butter. It grows slowly, always in troops, and is seldom attacked by maggots. Do not confuse it with False chanterelle (p 17), which has thin true gills and no smell. **Cap** 3–10 cm, soon depressed to funnel-shaped, with wavy margin, egg yellow to apricot. **Gills** decurrent, thick, formed by forking veins and ridges, widely spaced. **Stem** 2–5 cm, paler than cap, smooth. *July–Nov.*

## Amanita phalloides
### Death cap

*Streaky cap*

**Spores** 8–11 × 7–9 μm, amyloid

*White, free gills plus ring on stem*

*Large volva*

× 0.5

The most poisonous mushroom known, responsible for 90% of all deaths from mushroom poisoning; it is deadly even when cooked. Small amounts (20 g) are fatal and hands must always be washed after handling. About the same size as a Horse mushroom (p 153), but the cap colour, white gills and volva are distinctive. Gastro-enteric symptoms commence 6–15 hours after ingestion, with severe liver damage by the third day. **Cap** 6–15 cm, convex, moist, satiny, radially fibrillose, yellowish green to olive, discolouring to cream. **Gills** white, crowded. **Stem** 8–20 cm, cylindric with bulbous base cupped by fleshy, sac-like volva; white, slightly scaly, with thin, membranous, persistent ring. **Flesh** thick, white, with smell of honey, soon ammoniacal. *July–Nov.*

## A. citrina
### False death cap

**Spores** 8–11 × 7–9 μm, amyloid

*Swollen base with volva*

× 0.3

Best avoided for risk of confusion with Death cap. Smells of raw potatoes. **Cap** 4–8 cm, convex, with thin, non-striate margin, lemon yellow to almost white, with loose scales. **Gills** free, white. **Stem** 6–12 cm, white, with large basal bulb; ring striate. *Aug–Oct.*

## A. virosa
### Destroying angel

**Spores** 9–13 × 8–12 μm, amyloid

*Slender scaly stem supports conical cap*

*Thick, sac-like volva cups bulbous base*

× 0.3

Deadly, with a heavy soporific smell; do not mistake for *Agaricus silvicola* (p 35). **Cap** 6–10 cm, never depressed, white, sometimes pinkish. **Gills** free, white, crowded. **Stem** 8–15 cm, white, fibrillose; ring thin, fragile. *Aug–Oct.*

Beech and oak woods

## *Cantharellus infundibuliformis*

**Spores**
8–12 ×
5–10 μm,
ellipsoid

A common woodland species, always found in small groups, often in large numbers. Although it has a thinner flesh than Chanterelle (p 103), the flavour is excellent and ideal in stews. The gills are more strongly developed than in *C. lutescens* (p 66). **Cap** 2–5 cm, deeply funnel-shaped with wavy margin, dark brown. **Gills** formed by narrow veins and folds, yellow then greyish, decurrent, irregular. **Stem** 2–7 cm, deep yellow. *July–Dec.*

*Yellow compressed stem*

*Intervening gills*

× 0.5

*Often in clusters*

## *Craterellus cornucopioides*
### Horn of plenty

**Spores**
12–14 ×
7–8 μm,
ellipsoid

*Surface lacks gills*

Although unattractive and thin-fleshed, these are delicious to eat, described in France as "la viande des pauvres". They are usually cut into small pieces before cooking. Can grow in large troops, but the dark colour makes them difficult to see. **Cap** 4–8 cm, tubular, trumpet-shaped, lobed with thin margin, dark blackish brown to brownish grey, shiny when dry, fibrillose-scaly. **Underside** waxy, pale grey to bluish grey, powdery. **Stem** 4–12 cm, tubular, compressed, brownish black. *Aug–Oct.*

× 0.5

## *Collybia confluens*
### Clustered tough shank

**Spores**
7–9 ×
3–4 μm,
ellipsoid

Commonly found in beech woods, amongst dead leaves, always in clusters with the stems united at the base. The stems are covered in white down, giving a powdery appearance. **Cap** 2–4 cm, convex or campanulate, expanding, pinkish brown drying paler, thin, tough. **Gills** adnexed, pink to whitish, narrow, crowded. **Stem** 7–10 cm, pinkish or purplish, finely hairy, often compressed. *July–Nov.*

*Tufted, with stem bases joined*

× 0.5

## *Collybia dryophila*
### Russet shank

*Spores*
$4-7 \times 3-4\,\mu m$,
ellipsoid

× 0.6

Frequently occurs amongst dead leaves in oak woods, usually in small groups. The thin flesh is edible after cooking but is mildly poisonous if eaten raw. Also popularly known as "Penny top". **Cap** 2–5 cm, flattened, pale yellowish to reddish brown, with a brown centre, smooth. **Gills** adnexed, white or pale yellow, narrow, crowded. **Stem** 4–8 cm, slender, yellow to light reddish brown, hollow, smooth. *July–Nov*.

## *C. fusipes*
### Spindle shank

*Spores*
$4-6 \times 2-4\,\mu m$,
ellipsoid

A tough sp, found in dense tufts around the base of beech and oak. The hard swollen stems taper below into a common rooting base within the host tree. It decays slowly and old specimens are often mouldy. **Cap** 3–10 cm, broadly campanulate with incurved margin, tawny brown, paling from centre, shiny, smooth. **Gills** irregular, broad, white to grey, often rust-spotted. **Stem** 8–12 cm, compressed, grooved, white above, red-brown below. *July–Nov*.

× 0.3

## *Oudemansiella mucida*
### Slimy beech caps

*Slimy caps*

*Spores* very large,
$13-19\,\mu m$ diam

Only grows on beech, often in large numbers on stumps and over-hanging branches. The translucent, slimy cap is said to resemble a poached egg. **Cap** 3–8 cm, convex expanding, pure white to pale grey, very glutinous, becoming wrinkled as it dries. **Gills** adnate, white, broad, distant. **Stem** 3–8 cm, thin near apex, tough, white, often curved; ring high, pendent, grooved. **Flesh** white, soft, slimy. *Aug–Nov*.

*On dead trunks and branches of beech*

× 0.5

## *O. radicata*

### Rooting shank

*Spores 12–16 × 10–12 µm, ellipsoid*

A large, tall, solitary mushroom with a sticky cap, even in dry weather. It arises from a long, rooting base which is attached to buried wood, especially beech. Edible, but not recommended. **Cap** 4–10 cm, convex then flat, umbonate, wrinkled, yellowish to olive brown. **Gills** thick, white, adnexed, broad, distant. **Stem** 10–20 cm, twisted, cartilaginous, brown. **Flesh** thin. *July–Oct.*

× 0.3

## *Mycena inclinata*

*Spores 8–12 × 5–7 µm, ellipsoid, amyloid*

Dense clusters on oak stumps

May be distinguished from *M. galericulata* (p 68) by its reddish-brown stem base. It has a rancid or alkaline smell. **Cap** 2–4 cm, conical expanding, deep red-brown to grey, smooth, striate, with the margin extending beyond the gills. **Gills** adnate, white then pinkish, crowded. **Stem** 5–10 cm, tough, red-brown, white above. *Aug–Nov.*

× 0.5

## *M. vitilis*

*Spores 8–11 × 5–6 µm, ellipsoid, amyloid*

Grey-brown, striate cap

Long, shiny stem

A small species with a tall, slender stem; grows either singly or in small groups amongst dead leaves. **Cap** 1–3 cm, conical expanding, with a small, pointed umbo, greyish brown, striate. **Gills** adnexed, white, fairly crowded. **Stem** 7–15 cm, pale grey, smooth, shiny. *Aug–Dec.*

× 1

## Mycena pura
### Lilac mycena

**Spores** 7–8 × 3–5 μm, amyloid

Smell of radish

◀ Pink or lilac cap with striate margin

× 0.3

Do not confuse with *Laccaria amethystea* (p 66), which has thick, violet gills. The pink colour is often washed out and appears whitish. **Cap** 2–8 cm, campanulate expanding, rose pink to lilac. **Gills** adnate with decurrent tooth, interveined, white or pink, broad, spaced. **Stem** 4–8 cm, cap colour, smooth, base downy. *June–Dec.*

## M. pelianthina

**Spores** 5–7 × 2–3 μm, amyloid

× 0.3

Edge of gills dark violet

Grows amongst dead leaves near beech. Recognized by the dark gill edge. **Cap** 2–4 cm, convex or campanulate, purplish brown drying paler, translucent with a striate margin. **Gills** adnate to sinuate, greyish violet, spaced. **Stem** 5–8 cm, cap colour, striate. **Flesh** smells of radish. *Aug–Nov.*

## Hygrophorus eburneus
### Ivory slime cap

**Spores** 7–10 × 4–5 μm, oblong-ellipsoid

× 0.3

A very slimy, whitish sp, with thick, decurrent gills and a tapering stem base. Sometimes it is very abundant on the ground in oak woods. **Cap** 4–10 cm, fleshy, convex then flattened, with an incurved margin. **Gills** short-decurrent, thick, white, spaced. **Stem** 4–9 cm, white, with powdery apex and a tapering, rooting base. **Flesh** thick, white. *Aug–Oct.*

## H. chrysaspis

**Spores** 7–9 × 4–6 μm, ellipsoid

All surfaces discolour rusty brown

× 0.3

Similar to *H. eburneus* (left) but found under beech on chalky soil. Soon discolours rusty yellow or brown, with the gills blackening on drying. Smells of formic acid. **Cap** 3–6 cm, convex to umbonate, pure white at first, very slimy. **Gills** decurrent, creamy white turning blackish when dry, spaced. **Stem** 4–6 cm, tapering below, cream, becoming brownish towards base. *Aug–Oct.*

Beach and oak woods

## *Panellus serotinus*

**Spores** 4–7 × 1–2 µm, amyloid

*Flesh is soft and gelatinous*

× 0.3

Occurs in late autumn and early winter, attached to trunks and fallen logs by a short, lateral stem. More common in northern beech woods. **Cap** 3–8 cm, fan-shaped, convex, olive green, slimy, with inrolled margin. **Gills** adnate, yellow, narrow, crowded. **Stem** 1–2.5 cm, yellowish brown, with small, olive-brown scales. *Sept–Dec.*

## *Marasmius alliaceus*

**Spores** 9–10 × 6–8 µm, ellipsoid

*Strong smell of garlic*

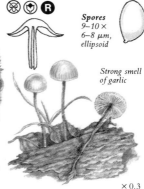

× 0.3

Although rare in Britain, it is locally common in the beech woods of northern Europe. The tall black stem has a fine, velvety surface. **Cap** 2–4 cm, campanulate expanding, whitish then greyish brown, with striate margin. **Gills** whitish, narrow, spaced. **Stem** 10–20 cm, ridged, rooting. *Aug–Nov.*

## *M. cohaerens*

**Spores** 8–10 × 2–5 µm, ellipsoid

× 0.5

Grows in troops amongst dead beech leaves. The finely velvety cap and tough, smooth stem are characteristic. **Cap** 1–3 cm, conical expanding, umbonate, pale cinnamon to tawny brown, minutely velvety. **Gills** adnexed, ochre yellow to cream, often with a brown edge, spaced. **Stem** 5–10 cm, pale to dark brown, smooth, shiny. *Oct–Feb.*

## *M. wynnei*

**Spores** 6–7 × 3–4 µm, ellipsoid

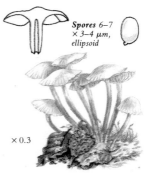

× 0.3

Also known as *M. globularis*. Always grows in clusters near beech, forming an extensive white mycelial mat which binds the dead leaves together. **Cap** 1–5 cm, hemispherical to convex, white becoming greyish violet, with a translucent margin. **Gills** adnexed, white, narrow, spaced. **Stem** 4–8 cm, white above, red-brown below. *Aug–Nov.*

## *Pleurotus ostreatus*
### Oyster mushroom

Spores $8-11 \times 3-4\ \mu m$, cylindric

Short stem
× 0.3

A well-known edible sp, grown commercially in Hungary. Forms clusters on frondose trees, esp beech. Best eaten when young. **Cap** 7–14 cm, excentric, convex, ash grey, black, purplish or white, smooth, shiny. **Gills** decurrent, white to cream, very crowded. **Stem** 1–3 cm, lateral, solid, hard. *Jan–Dec.*

## *P. cornucopiae*
### Branched oyster fungus

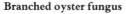
Spores $7-11 \times 3-5\ \mu m$, cylindric

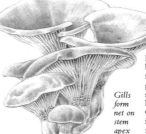
Gills form net on stem apex
× 0.3

In dense tufts, usually on beech, but also on other trees. Differs from *P. ostreatus* (above) by its gills, which branch at the point of stem attachment. Very young fruit-bodies are edible after cooking. **Cap** 7–10 cm, excentric, convex or infundibuliform with inrolled margin, pale grey-brown to white, shiny, fibrillose. **Gills** whitish, decurrent. **Stem** 1–2 cm, lateral or central, whitish. *June–Sept.*

## *Schizophyllum commune*
### Split gill

Section shows gills splitting

Spores $3-4 \times 1-2\ \mu m$, cylindric

× 0.6
Tough, dry brackets grow on dead wood

A tough, dry fungus which can withstand periods of drought. Found occasionally in southern England, but not further north; much more common in central Europe. It is popular as a form of chewing gum in tropical Africa. **Cap** 1–3 cm, fan-shaped, greyish brown drying whitish, covered with small, pointed scales; margin incurved, lobed and irregular. **Gills** grey-brown or purplish, radiating, on drying each gill divides along its edge and rolls back. **Stem** short or absent, lateral. *May–Dec.*

Beech and oak woods

## *Tricholoma argyraceum*

**Spores**
$5-6 \times 3-4\ \mu m$, ovoid

One of a group of *Tricholoma* spp that have scaly caps; also known as *T. scalpturatum*. It is similar to the more common *T. terreum* (p 132), but it has paler gills, which turn yellow with age on the margin. Found amongst beech litter, especially on chalky soil. Edible, with a mealy taste and smell. **Cap** 3–7 cm, convex, umbonate, mouse grey, with numerous darker fibrillose scales. **Gills** sinuate, pale grey, drying bright yellow at the cap margin, crowded. **Stem** 4–7 cm, white or with greyish tints, without a ring. **Flesh** greyish white. *Aug–Nov.*

Gills stain yellow

On beech litter

× 0.6

## *T. columbetta*

**Spores**
$5-7 \times 4-5\ \mu m$, broadly ellipsoid

Satiny white cap surface

A large and particularly beautiful mushroom with a satiny white surface, which often discolours with pink or bluish spots. Not common in Britain but much more frequent in northern Europe, under beech and sometimes birch. A good edible fungus, if rather fibrous, with a sweet taste but no smell. Sold in market places in Switzerland. **Cap** 5–10 cm, convex expanding, pure white, spotting pink or blue, dry, shiny, sticky when moist. **Gills** sinuate, white, broad. **Stem** 6–8 cm, robust, cylindric, hard, solid, white discolouring pink or blue-green at base, fibrillose. **Flesh** thick, firm. *Aug–Nov.*

Broad white gills

Rooting stem base

× 0.5

111

## *Russula aurata*

**Spores**
$7–10 \times 6–8$ μm, ovoid, with spines and crests, amyloid

An excellent edible mushroom, found in damp places near oak, beech and hazel. **Cap** 5–10 cm, convex to depressed, golden yellow with pink tints near margin, often red at centre, moist, shiny. **Gills** adnate, broad, interveined, fairly crowded, cream to ochre. **Stem** 6–9 cm, soon hollow, white with lemon-yellow tints. **Flesh** fragile, white, mild. *Aug–Oct.*

× 0.5

## *R. lepida*

**Spores**
$8–9 \times 7–8$ μm, almost spherical, warty, amyloid

*Cap with velvety bloom*

A beautiful, pale red species, sometimes discolouring yellowish white. The hard white flesh has a faint odour of menthol or cedar-wood and a bitter taste. **Cap** 3–8 cm, convex expanding, vermilion red at centre, paler near margin, often pinkish, dry. **Gills** adnate, crowded, cream. **Stem** 3–10 cm, white flushed pink. *June–Nov.*

× 0.6

## *R. mairei*

*Cap with thick, hard flesh*

**Spores**
$7–8 \times 6–7$ μm, ovoid, with ridges, amyloid

Resembles *R. emetica* (p 25) and will also grow under conifers, but its flesh is hard and firm (not thin and fragile) with a smell of coconuts when young. Typically associated with beech, even grows under isolated trees. It has a very acrid taste. **Cap** 4–7 cm, strongly convex, finally slightly depressed, scarlet to blood red. **Gills** adnate, whitish, crowded. **Stem** 2–5 cm, cylindric, hard, solid, white. *Sept–Oct.*

× 0.6

Beech and oak woods

## *R. fellea*
### Geranium-scented russula

**Spores** $7–9 \times 6–7\ \mu m$, ovoid, network, amyloid

A late autumnal species, uniformly coloured and with a distinctive smell of geraniums. Commonly found under beech and oak. **Cap** 3–8 cm, convex expanding, soon depressed, pale ochre to tawny ochre, sticky, shiny. **Gills** adnate, ochre to cream, fairly crowded. **Stem** 3–7 cm, variable, solid, pale yellowish brown. **Flesh** thick, white, with very acrid taste. *Sept–Nov.*

× 0.6

*All parts uniformly straw-coloured*

## *R. parazurea*

**Spores** $6–9 \times 5–7\ \mu m$, ovoid, partial net, amyloid

This species is more frequent in Britain than in other parts of northern Europe. **Cap** 5–7 cm, convex to depressed, dark blue-green to olive grey, often with a powdery bloom. **Gills** adnate, white to cream, fairly crowded. **Stem** 3–4 cm, white, soon hollow. **Flesh** white, mild. *June–Sept.*

× 0.6

*Mild or slightly hot taste*

## *R. atropurpurea*
### Blackish-purple russula

**Spores** $7–9 \times 6–7\ \mu m$, ovoid, network, amyloid

The flesh of this mushroom is very firm with a vague smell of apples, but is inedible because of its acrid taste. **Cap** 5–12 cm, convex expanding, irregular, purplish red to almost black at the centre, sometimes developing creamy-yellow patches. **Gills** adnate, cream, fairly crowded. **Stem** 5–6 cm, cylindric, white. **Flesh** white to greyish. *June–Oct.*

× 0.6

Beech and oak woods

## *Russula pseudointegra*

*Spores* 7–9 × 6–8 μm, almost spherical, warty, amyloid

The white stem and ochre gills separate this sp from *R. lepida* (p 112). Found under oak, beech and hazel. **Cap** 5–12 cm, strongly convex, expanding, cinnabar red to scarlet, discolouring pale pink; margin striate. **Gills** adnate, whitish then ochre, fairly crowded. **Stem** 4–8 cm, cylindric, pure white. **Flesh** hard, bitter. *July–Oct.*

× 0.3

## *R. virescens*

White stem

*Spores* 7–9 × 6–7 μm, ovoid, warty, amyloid

Regarded as one of the best edible *Russula* spp with a nutty flavour; can be eaten raw. Often in grass in small troops, esp near beech. **Cap** 6–12 cm, convex expanding to depressed, grey-green, blue-green, olive to brownish, dry, cracking into small scales. **Gills** creamy white, fairly crowded. **Stem** 4–8 cm, cylindric, solid. **Flesh** thick, with fruity smell. *June–Oct.*

× 0.5

## *Lactarius volemus*

× 0.3

*Spores* 8–10 μm diam, spherical, amyloid

Young specimens have a sweet honey smell which becomes fishy with age. **Cap** 5–15 cm, convex to depressed, reddish golden to tawny orange, becoming paler, velvety. **Gills** adnate, cream then orange-yellow bruising dark brown, forked, fairly crowded. **Stem** 5–10 cm, cap colour, powdery. **Flesh** cream, tawny brown when cut; milk sticky, white, mild. *Aug–Oct.*

Beech and oak woods

## *L. blennius*

### Slimy milk cap

*Slimy cap*

**Spores**
$6–9 \times 5–7$ μm,
ridged, amyloid

Always under beech, this sp may be recognized by its slimy, grey-green cap, which is concentrically zoned with dark drop-like markings. It produces a very hot white milk, which turns grey on exposure. **Cap** 5–10 cm, convex to depressed, olive brown to green-grey. **Gills** white staining grey, slightly decurrent. **Stem** 3–5 cm, tapers below, paler than cap. **Flesh** white. *Aug–Nov*.

× 0.5

## *L. quietus*

### Oak milk cap

**Spores**
$8–9 \times 7–8$ μm,
spiny, ridged, amyloid

Restricted to oaks, its zoned cap and strong oily odour make it easy to recognize. **Cap** 4–10 cm, convex to depressed, dull reddish brown, distinctly zoned with darker spots. **Gills** slightly decurrent, white to reddish, crowded. **Stem** 3–7 cm, cap colour or darker, furrowed; milk white, mild. *Sept–Nov*.

*Oily smell*

× 0.6

## *L. pallidus*

*Flesh produces mild white milk*

**Spores**
$8–10 \times 6–7$ μm,
ovoid, with short crests, amyloid

Usually found under beech, but may also occur with birch; it is often half-hidden in the leaf litter, sometimes in large quantities. Look out for its uniformly pale, smooth cap with inrolled margin. **Cap** 6–10 cm, convex to slightly depressed, pale pinkish brown, sticky. **Gills** slightly decurrent, paler than cap, crowded, narrow. **Stem** 5–7 cm, cylindric, smooth, cap colour or paler. *July–Sept*.

× 0.5

Beech and oak woods

## *Phylloporus rhodoxanthus*

*Dry, velvety cap*

*Thick, waxy, interveining gills*

**Spores**
$10–13 \times 3–5$ μm, spindle-shaped

A fairly rare sp, growing either singly or in small groups, often on sandy banks or in poor soil. It closely resembles *Xerocomus subtomentosus* (p 61), but it has thick, waxy, branching gills instead of pores. **Cap** 2–9 cm, convex expanding, finally depressed, yellowish brown, almost felty, dry, often cracked. **Gills** adnate-decurrent, forking, interveined, spaced. **Stem** 2–10 cm, tapering below, solid, reddish brown, ridged. **Flesh** thick, yellowish. *July–Oct*.

*Tapering stem*

× 0.5

## *Cortinarius hinnuleus*

**Spores**
$8–9 \times 5–6$ μm, ovoid, warty

Found in mixed woods, but especially under oak; it is recognized by its silky cap, widely spaced gills and the flaky white ring zone on its stem. *C. gentilis* (p 32) is similar but has yellow zones on the stem. *C. hinnuleus* is one of the earliest *Cortinarius* spp to appear in autumn and has an earthy smell. **Cap** 2–6 cm, flat, umbonate, yellow-brown, smooth, with a split margin. **Gills** adnate, with interveining, ochre brown to rust brown, spaced. **Stem** 5–7 cm, fibrillose, yellowish brown, with white ring zone. *Aug–Nov*.

× 0.75

Beech and oak woods

## *C. elatior*

**Spores**
12–17 × 7–9 μm, almond-shaped, warty

Grows under beech, although it may also be found in coniferous woods. Both the wrinkled cap and spindle-shaped stem are slimy. **Cap** 5–12 cm, conical to convex, olive brown sometimes with a violet margin, wrinkled. **Gills** adnate, clay brown with violet tint, finally rusty brown, broad, fairly crowded. **Stem** 6–18 cm, tapers below, pale violet, silky-striate, slimy. *Aug–Nov*.

*Violet gills soon dark brown*

× 0.3

## *C. pseudosalor*

**Spores**
12–14 × 7–9 μm, warty

Variable in colour, but typified by the smooth white stem with blue-violet tints, and the slight blue tinge on the gills. Paler and smaller than *C. elatior* (above), growing mostly under beech or hazel, sometimes birch. **Cap** 3–7 cm, conical to convex, ochre yellow to buff. **Gills** adnate-decurrent, crowded, ochre brown. **Stem** 8–10 cm, tapering. *Aug–Nov*.

× 0.5

## *C. caesiocyaneus*

**Spores**
8–10 × 4–5 μm, almond-shaped, warty

Prefers chalky soil in beech woods. The stem is characterized by a wide bulbous base with a distinct rim and an abundant bluish cortina. Has an unpleasant, musty smell. **Cap** 3–8 cm, at first pale blue, discolouring ochre from centre, sticky, fibrillose. **Gills** adnate, cap colour, edge remaining blue. **Stem** 3–7 cm, thick, cap colour becoming ochre. *Sept–Oct*.

*Bulbous base with rim*

× 0.5

117

Beech and oak woods

## *Cortinarius melliolens*

Smell of honey

**Spores**
8–9 ×
4–5 μm, warty

×0.5

Recognized by its brightly coloured, sticky cap and smell of honey. Usually associated with beech, but may also grow under conifers and birch. **Cap** 4–10 cm, convex expanding, bright ochre yellow, with a whitish veil. **Gills** adnate, broad, clay to rusty brown. **Stem** 5–8 cm, club-shaped with a basal bulb, white to ochre yellow with a white cortina. *Aug–Nov*.

## *C. torvus*
### Sheathed cortinarius

Dry cap

**Spores**
8–10 ×
5–6 μm,
ellipsoid,
warty

A large, fleshy sp, often under beech. Edible, but of poor quality, with a smell of camphor. The violet tints fade with age. **Cap** 4–10 cm, convex, fleshy, dark brown, smooth. **Gills** adnate, at first violet then cinnamon brown, thick, spaced. **Stem** 7–10 cm, pale violet at apex, surrounded by a whitish to violet-brown sheath, which forms a persistent membranous ring; swollen, tapers towards base. *Aug–Nov*.

×0.5

## *C. anomalus*

Gills turn rusty brown

**Spores** 8–10 ×
6–8 μm, warty

Mainly under beech, but will also grow under oak and birch. Has a slender stem and blue-violet gills, which turn rusty brown. **Cap** 3–7 cm, fleshy, convex expanding, umbonate, pale grey then greyish yellow, silky. **Gills** adnate, crowded. **Stem** 6–10 cm, cylindric, at first white with a blue apex, drying pale ochre, with yellowish patches left by the veil. *Aug–Oct*.

×0.5

Beech and oak woods

## *Hebeloma radicosum*

**Spores**
$8–10 \times 4–6\ \mu m$,
almond-shaped, warty

Sticky cap

× 0.3

Tends to grow around old stumps. This sp is not typical of its genus in having a persistent membranous ring on the stem and a long rooting base, and is often mistaken for a *Pholiota* sp. Edible but poor. **Cap** 6–9 cm, convex, pale brown, with small white scales. **Gills** sinuate, cinnamon brown, crowded. **Stem** 5–8 cm, whitish, fibrillose-scaly; ring double. *Aug–Nov*.

## *Inocybe fastigiata*     *I. maculata*

**Spores**
$8–14 \times 4–7\ \mu m$,
bean-shaped

Conical cap

× 0.5

A very common *Inocybe* sp, with a distinctive conical, yellowish-brown cap and a smell of mouldy bread or soap. Causes muscarine poisoning and should be avoided. Usually grows under beech. **Cap** 3–8 cm, campanulate, umbonate, yellowish brown, fibrillose, cracking towards the margin. **Gills** adnate, olive yellow becoming olive brown, rather narrow, crowded. **Stem** 4–10 cm, whitish to ochre yellow, fibrillose, hollow. *June–Oct*.

**Spores**
$8–12 \times 4–8\ \mu m$,
bean-shaped

Cracked cap

× 0.5

Closely related to *I. fastigiata* (left), but with darker brown fibrils on the cap and stem. Also contains muscarine compounds and therefore should not be eaten. **Cap** 2–8 cm, campanulate with a pointed umbo, chestnut brown, radially cracking, at first covered by a fine white veil. **Gills** adnate, whitish, finally snuff brown, crowded. **Stem** 4–9 cm, fibrillose-striate, white then brown, with a bulbous base. **Flesh** whitish, with a slightly aromatic smell. *June–Oct*.

Beech and oak woods

## *Coprinus picaceus*
### Magpie ink cap

*Spores 13–17 × 10–12 µm, ellipsoid, with germ-pore*

Large white patches on cap

Black, deliquescent gills

A robust ink cap, easily identified by its black and white colouring, which has given rise to its common name "Magpie". Usually found solitary on the ground in beech woods. **Cap** 4–7 cm, ovoid to conical, covered when young by a white felty veil, which cracks as the cap expands leaving large white patches on a dark brown to black background. **Gills** white then black, crowded, deliquescent. **Stem** 12–20 cm, white, fragile, hollow, smooth. *Sept–Dec.*

× 0.3

## *Psathyrella obtusata*

*Spores 7–9 × 4–5 µm, ellipsoid, with germ-pore*

Striate cap margin

× 0.75

Very common on the ground in shady woodland, especially under oak. **Cap** 1–2 cm, convex then flattened, date brown drying out from the centre to pale tan, wrinkled, with margin remaining dark and striate. **Gills** adnate, pale to dark brown, rather spaced. **Stem** 6–8 cm, white, silky, shiny, fragile, hollow. *Apr–Nov.*

## *Psilocybe squamosa*

*Spores 11–14 × 6–7 µm, ellipsoid, with germ-pore*

× 0.5

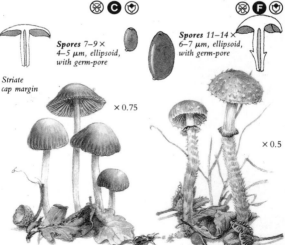

Not common but can occur in very large numbers on the ground in beech woods, attached to buried twigs or beech-mast. **Cap** 2–4 cm, hemispherical expanding, ochre yellow, sticky when moist, with scattered small white scales. **Gills** adnate, dark grey-brown with white edge, crowded. **Stem** 7–10 cm, white above ring, rusty brown below, scaly. *Sept–Nov.*

Beech and oak woods

## *Pholiota adiposa*

**Spores**
$5-7 \times 3-4$ μm,
ellipsoid, with
germ-pore

Sticky scales
on cap and stem

Found only in association with
beech trees, where it forms clusters
on old roots. Similar to *P. aurivella*
(p 90), but it has sticky gelatinous
scales on both the stem and cap.
Edible but not recommended. **Cap**
6–12 cm, fleshy, convex expanding,
golden yellow, with rusty-brown
gelatinous scales arranged in
concentric zones. **Gills** sinuate,
rusty yellow, broad. **Stem** 6–15 cm,
cylindric, yellowish brown, darker
towards base, covered with
gelatinous scales; ring fibrillose,
indistinct. *July–Sept.*

× 0.3

## *Boletus appendiculatus*

Dry, velvety
cap

**Spores** $12-15 \times 4-5$ μm,
spindle-shaped

Sulphur-
yellow
pores

Related to the Penny bun boletus
(p 123) and is equally as edible
and delicious. Occurs in early
summer and is gone by autumn,
preferring warmer southern areas,
where it grows singly or in small
groups. Recognized by the swollen
subrooting stem and the sulphur-
yellow pores. **Cap** 8–20 cm, hemispherical to convex, dry, with
fine cracks, yellowish brown to
dark brown. **Tubes** adnate, lemon
yellow to greenish, depressed
around stem. **Pores** small, round,
bluing slightly. **Stem** 7–12 cm,
robust, solid, sulphur yellow,
brownish at base, with hexagonal
network on upper surface. **Flesh**
thick, pale yellow, not or hardly
bluing on exposure. *July–Sept.*

× 0.5

Beech and oak woods

## *Boletus satanas*
# The devil's boletus

*Flesh turns blue when moist*

*Spores 11–14 × 5–6 μm, spindle-shaped*

One of the largest boletes, found only occasionally in late summer to early autumn. It causes severe gastro-intestinal poisoning, even in small amounts, but it is not fatal. **Cap** 8–30 cm, hemispherical to strongly convex, with thick margin, whitish, at times greyish or olive brown, shiny, finally cracking. **Tubes** adnate, short, yellow, bluing when cut. **Pores** round, yellow to brick red, paler with age, bruising blackish blue. **Stem** 5–10 cm, very swollen, upper region yellowish with fine but distinct red network, carmine red elsewhere, bluing to the touch. **Flesh** yellowish white, bluing when moist, with an unpleasant smell. *July–Oct.*

*Solitary, on chalky soil*

× 0.3

## *B. luridus*

*Yellowish-olive cap*

*Flesh yellow, turns deep blue when broken*

*Spores 11–13 × 5–7 μm, spindle-shaped*

Much more common than *B. satanas* (above), with which it is often confused, and similarly found on chalky soil under beech, oak or lime. Poisonous when eaten raw yet edible when cooked, although it discolours deep blue. Note the red pores and the elongated red network on the stem. **Cap** 6–20 cm, hemispherical to convex, matt, yellowish olive at centre, elsewhere orange to brick red. **Tubes** depressed around the stem, olive yellow, bluing. **Pores** round, yellow, blood red to orange, bruising blue. **Stem** 8–20 cm, solid, often bulbous, with red network. *May–Dec.*

× 0.3

Beech and oak woods

## *B. edulis*
### Penny bun boletus, Cep

**Spores**
*16–19 ×
5–6 µm,
spindle-shaped*

*Cap sticky
when moist*

*Upper stem
covered with
fine network
pattern*

Known in France and Italy as the "King of Mushrooms" because of its delicious, white, nut-flavoured flesh. Do not confuse with *Tylopilus felleus* (p 124). **Cap** 8–30 cm, hemispherical to convex with thick margin, whitish to blackish brown, often coffee-coloured or chestnut brown. **Tubes** free, white then greenish yellow. **Pores** small, round, white to olive yellow, not bluing. **Stem** 10–20 cm, hard, pale brown, always thick. **Flesh** white, unchanging. *June–Nov.*

*Stem base
whitish,
swollen*

× 0.3

## *B. aestivalis*

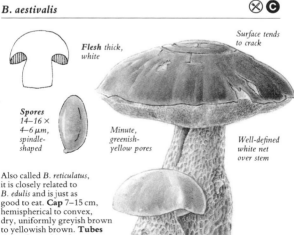

**Flesh** *thick,
white*

**Spores**
*14–16 ×
4–6 µm,
spindle-shaped*

*Minute,
greenish-
yellow pores*

*Surface tends
to crack*

*Well-defined
white net
over stem*

Also called *B. reticulatus*, it is closely related to *B. edulis* and is just as good to eat. **Cap** 7–15 cm, hemispherical to convex, dry, uniformly greyish brown to yellowish brown. **Tubes** short, fine, white then greenish yellow. **Pores** white then olive green. **Stem** 5–15 cm, robust, swollen, solid, whitish ochre then pale brown, paler than cap, with white network. *May–Oct.*

× 0.3

## *Gyroporus castaneus*

## *G. cyanescens*

**Spores** 8–13 × 5–8 μm, ellipsoid

*Tough stem*

× 0.5

**Spores** 8–11 × 5–6 μm, short-ellipsoid

× 0.5

An excellent edible fungus, after the stems have been discarded, but not very common; prefers beech, oak and chestnut woods. **Cap** 3–10 cm, hemispherical to convex, often depressed, dry, chestnut brown, reddish brown to cinnamon. **Tubes** depressed around the stem, short. **Pores** round, white then lemon yellow, bruising brown. **Stem** 4–7 cm, spindle-shaped, finally hollow, paler than cap, ridged. **Flesh** white, unchanging or pinkish when cut. *Aug–Oct.*

Prefers stony soil; also found on heaths with spruce and birch. Do not be discouraged by the blue discoloration of the white flesh, as it is very good to eat. **Cap** 3–12 cm, off-white to straw yellow, bruising brown, velvety-scaly. **Tubes** white then straw yellow, bruising blue. **Pores** small, round, white then yellow, bluing. **Stem** 5–10 cm, cap colour, often cracking to form ring zones, soon hollow. *July–Nov.*

## *Tylopilus felleus*
### Bitter boletus

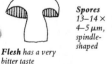

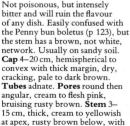

*Brown net on stem*

× 0.3

**Spores** 13–14 × 4–5 μm, spindle-shaped

*Flesh has a very bitter taste*

Not poisonous, but intensely bitter and will ruin the flavour of any dish. Easily confused with the Penny bun boletus (p 123), but the stem has a brown, not white, network. Usually on sandy soil. **Cap** 4–20 cm, hemispherical to convex with thick margin, dry, cracking, pale to dark brown. **Tubes** adnate. **Pores** round then angular, cream to flesh pink, bruising rusty brown. **Stem** 3–15 cm, thick, cream to yellowish at apex, rusty brown below, with ridged brownish veins. **Flesh** white, unchanging, with intensely bitter taste. *June–Nov.*

# Alder carrs

A copse of alder trees growing in a boggy area with base-rich soil is known as an alder carr. Such a wood is typical of marshes and fenland, and was once widespread throughout northern Europe before the introduction of land-drainage schemes. The alder tree is able to withstand the lack of oxygen in waterlogged soil by obtaining its nitrogen, so essential for growth, via nitrogen-fixing bacteria present in the root nodules. The typical humps and tussocks that flourish at the tree base provide a habitat of relative dryness suitable for an abundant fungus flora, which the rich soil is able to support, although it is not clear whether the fungi simply prefer a habitat similar to alder or whether there is a more intimate relationship with the tree.

## *Mycena tortuosa*      *Lactarius obscuratus*

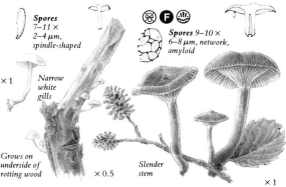

**Spores** 7–11 × 2–4 µm, spindle-shaped

× 1

Narrow white gills

Grows on underside of rotting wood  × 0.5

**Spores** 9–10 × 6–8 µm, network, amyloid

Slender stem

× 1

Easily overlooked, this minute sp is always found in wet places on the underside of fallen logs, often alder but also other trees in moist situations. Both the cap and stem are covered in very fine hairs. **Cap** 1–8 mm, small, conical to convex expanding, pure white, finely hairy. **Gills** adnate, fairly crowded. **Stem** 1–3 cm, short, slender, white, finely hairy as cap. *July–Nov.*

Too thin to be of culinary value, the flesh produces a white, watery milk with a mild to slightly acrid taste. **Cap** 1–2 cm, umbilicate with central papilla, orange-brown to buff, often olive green at centre, with thin, translucent, striate margin. **Gills** decurrent, ochre yellow, crowded. **Stem** 1–3 cm, cylindric, whitish to yellow-brown. *Aug–Oct.*

## Naucoria escharoides

**Spores** 10–11 × 5–6 μm.

## Gyrodon lividus

**Spores** 5–6 × 4–5 μm, ovoid

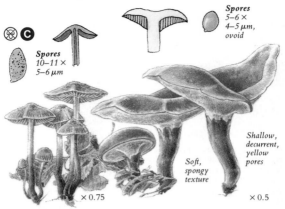

Shallow, decurrent, yellow pores

Soft, spongy texture

× 0.75

× 0.5

Often in large numbers. **Cap** 1–3 cm, convex expanding, yellowish to reddish brown, scurfy. **Gills** adnate, brown, narrow, crowded. **Stem** 3–5 cm, fibrillose, yellowish brown. *Aug–Oct*.

Known only from the Cambridgeshire fens in Britain, but more frequent elsewhere in northern Europe. Grows in groups near alder. **Cap** 3–10 cm, convex to depressed, irregular, sticky when wet, shiny when dry, olive beige to reddish brown, fibrillose, with thin, wavy margin. **Tubes** short, decurrent, lemon yellow. **Pores** ochre yellow to olive, bluing to touch. **Stem** 4–10 cm, curved, cap colour, fibrillose. **Flesh** thin. *Aug–Oct*.

## N. scolecina

**Spores** 9–11 × 5–6 μm

## Pholiota alnicola

**Spores** 8–10 × 4–6 μm, with germ-pore

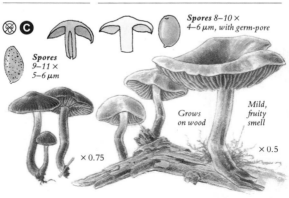

Grows on wood

Mild, fruity smell

× 0.75

× 0.5

Redder, less silky than *N. escharoides* (above). **Cap** 1–2 cm, convex, rusty brown then paler; margin striate. **Gills** adnate, pinky brown. **Stem** 3–7 cm, thin, curved. *Sept–Oct*.

Usually in tufts on old roots and fallen trunks of alder and willow in boggy ground. This is one of the few *Pholiota* spp which lacks a ring on the stem. It has a smell reminiscent of pear drops. **Cap** 2–8 cm, fleshy, convex expanding, yellow with olive or brownish tints; margin inrolled. **Gills** adnate to sinuate, straw yellow to cinnamon brown, quite crowded. **Stem** 4–8 cm, fibrillose, lemon yellow above, rust brown below. *Sept–Nov*.

# Elm

Until recently the elm used to be one of the most common English trees, its upright massive trunk being a characteristic feature of hedgerows and roadsides. Unfortunately, the Dutch elm disease, a microscopic fungus carried by the elm bark beetle, has drastically reduced the elm population in England from an estimated 23 million to a mere 6 million and sanitation felling, to contain the disease, has now been abandoned. The elms rarely occur in woodlands and reproduce mainly by sending out root suckers from old trees. Their trunks are notoriously prone to heart-rot, caused by several bracket fungi, so that apparently healthy, well-established trees nearly always carry a quantity of dead, rotting wood, offering a suitable substrate for certain mushrooms and toadstools.

## *Lyophyllum ulmarium*

**Spores** 5–6 μm diam, spherical

Adnate, crowded gills

The young specimens are edible although they taste slightly bitter, but the older fruitbodies are much too tough. The species grows in clusters on elm trunks, with several fruitbodies arising from a common base. Sometimes the cap is excentrically positioned on the stem, so that it was once regarded as a *Pleurotus* species. **Cap** 7–20 cm, convex expanding to flat, fleshy, dirty yellowish, smooth or cracked. **Gills** adnate, pale ochre yellow, crowded. **Stem** 8–12 cm, thick, whitish, firm, often curved upwards. **Flesh** white, tough. *June–Dec.*

× 0.5

Elm

## *Rhodotus palmatus*

*Thick, gelatinous pellicle becomes wrinkled with age*

**Spores**
*6–7 μm diam, spherical, warty*

*Gills sinuate, buff to reddish in colour*

*Found clustered on old stumps and fallen elm trunks*

× 0.6

Widespread but uncommon, sometimes found on old posts and beams. A pink-spored fungus, formerly included in many genera, but now thought distinct enough to warrant its own genus. The thick, firm flesh is edible but mediocre with a bitter taste. **Cap** 4–8 cm, pink to pale orange-buff with tough, wrinkled pellicle. **Gills** crowded. **Stem** 2–5 cm, white or reddish, excentric, fibrillose-striate. *Sept–Dec.*

## *Volvariella bombycina*
### Silky volvaria

**Spores**
*7–9 × 5–6 μm, ellipsoid*

*Cap covered with silky-fibrillose scales*

A fairly rare sp, this is the largest and surely the most beautiful of all the *Volvariella* spp. The fibrous-scaly cap is pale yellow at first but changes to white, and the thin flesh has a woody smell. The large white volva may lead to confusion with *Amanita* spp, but there is no ring and the gills are pink. Can also grow on trees other than elm. An edible fungus, but only of moderate quality. **Cap** 5–20 cm, parabolic, dry, shiny, slightly yellowish at centre with silvery margin. **Gills** white then pinkish brown, crowded. **Stem** 7–15 cm, satiny white, smooth. *June–Sept.*

*Emerges from large, white volva*

*Grows on rotten stumps and sawdust*

× 0.5

# Parks, roadsides and wasteland

This section covers those fungi that proliferate in areas where the soil has been artificially disturbed by humans. Thus the soil is usually well drained and rich in humus, providing ideal conditions for rapid colonization by weeds, including the "pioneer" mushrooms. These are the quickly growing, non-mycorrhizal species, such as some of the ink caps (*Coprinus*) and also some of the true mushrooms (*Agaricus*). The fast-growing underground mycelium is free from competition and radiates outwards, often producing either fairy rings, as in the Fairy ring champignon (p 131) and the Shaggy parasol (p 133), or clusters of fruitbodies as in *Lyophyllum connatum* (below). Roadside verges, where moving air currents rapidly disperse the spores, often prove to be a most rewarding site for mushroom hunters.

## *Lyophyllum connatum*

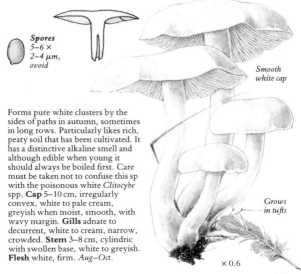

**Spores** 5–6 × 2–4 µm, ovoid

Forms pure white clusters by the sides of paths in autumn, sometimes in long rows. Particularly likes rich, peaty soil that has been cultivated. It has a distinctive alkaline smell and although edible when young it should always be boiled first. Care must be taken not to confuse this sp with the poisonous white *Clitocybe* spp. **Cap** 5–10 cm, irregularly convex, white to pale cream, greyish when moist, smooth, with wavy margin. **Gills** adnate to decurrent, white to cream, narrow, crowded. **Stem** 3–8 cm, cylindric with swollen base, white to greyish. **Flesh** white, firm. *Aug–Oct*.

Smooth white cap

Grows in tufts

× 0.6

## *Lyophyllum decastes*

**Spores** 5–7 μm diam, almost spherical

One of a group of closely related spp; also known as *Tricholoma aggregatum*. The dense clusters occur around stumps or on buried roots from summer onwards. The tough flesh is edible but may cause a burning sensation. **Cap** 4–12 cm, strongly convex finally expanding, umber to reddish brown, smooth. **Gills** adnate, white to yellowish, thin, crowded. **Stem** 5–10 cm, whitish to greyish brown, solid, fibrous. **Flesh** white. *July–Oct.*

× 0.5

## *L. fumosum*

**Spores** 5–7 × 4–6 μm, almost spherical

Differs from *L. decastes* in having greyish gills and stems that join together in a solid root-like base. Also known as *Tricholoma cinerascens*. Likes compost heaps and rich soil. **Cap** 5–10 cm, convex expanding, livid brown when moist, drying paler, smooth and shiny. **Gills** adnate, white soon greyish, crowded. **Stem** 7–8 cm, yellowish grey, solid. *July–Oct.*

× 0.5

## *Melanoleuca brevipes*

**Spores** 8–11 × 4–6 μm, warty, amyloid

Found singly or in groups, often by cinder paths. Similar to the more common *M. melaleuca* (p 75), but the stem is dark brown and shorter than the cap diameter. **Cap** 5–8 cm, convex then flat, umber brown becoming paler, smooth, moist. **Gills** sinuate, white, crowded. **Stem** 2–4 cm, fibrillose. *June–Oct.*

× 0.5

## *Leucopaxillus giganteus*
### Giant clitocybe

**Spores**
$6–8 \times 3–4\ \mu m$, ovoid, amyloid

A large white mushroom, exceptionally up to 45 cm tall, which often forms fairy rings in grassy places in parks and gardens. Said by some to be edible, but others are allergic to it and therefore is best avoided. Could be confused with *Clitocybe geotropa* (p 64) or *Lactarius vellereus* (p 77). **Cap** 15–30 cm, fleshy, convex soon depressed, cyathiform, creamy white, sometimes with silky scales at centre, with thin margin at first incurved. **Gills** decurrent, white to pale brown, very crowded. **Stem** 3–10 cm, stout, solid, whitish, fibrous. **Flesh** white, firm. *July–Sept.*

*Grows in groups or rings* ×0.3

## *Marasmius oreades*
### Fairy ring champignon

*Grows in large numbers forming fairy rings*

**Spores**
$7–10 \times 4–6\ \mu m$, ovoid

Much disliked by gardeners because it occurs in large numbers on lawns and at the edge of paths, especially during damp summers. It is a tough fungus and can survive periods of drought. A well-known edible sp, excellent when fried, but the tough stems are usually discarded. Do not mistake it for the poisonous *Clitocybe dealbata* and *C. rivulosa* (p 143), which grow in the same localities. **Cap** 2–5 cm, campanulate expanding, rusty brown, moist, paling from the margin to ochre yellow. **Gills** adnexed, whitish, widely spaced. **Stem** 2–7 cm, slender, colour as cap, fibrous. **Flesh** thin, fibrous. *May–Nov.*

×1

Parks, roadsides and wasteland

## *Lepista saeva*
### Blewit

*Flesh thick, white*

*Violet stem*

**Spores** $7–9 \times 4–5\ \mu m$, ellipsoid, warty

× 0.5

One of the best edible spp, superior to the Wood blewit (p 51), but not easily found. Likes chalky soil and may form large rings. Can tolerate very cold weather. **Cap** 5–12 cm, fleshy, convex expanding, dull grey-brown, smooth. **Gills** sinuate, grey to whitish, thin, crowded. **Stem** 5–10 cm, short, stocky, violet to mauve, fibrous-scaly. *Nov–Feb*.

## *L. sordida*

**Spores** $6–7 \times 3–4\ \mu m$, ellipsoid, warty

× 0.5

The grey or lilac flesh has a watery appearance and is thinner than other *Lepista* spp. Can occur in very large numbers on compost heaps and is often found under hedgerows. **Cap** 2–7 cm, convex soon flat or depressed, greyish lilac to brownish violet, smooth, drying paler. **Gills** sinuate, lilac to grey-brown, fairly crowded. **Stem** 4–6 cm, often curved, cap colour, fibrillose. *July–Nov*.

## *Tricholoma terreum*

**Spores** $6–7 \times 3–5\ \mu m$, ellipsoid

*Dark grey scaly cap*

× 0.5

A very common, widespread sp, found at the edge of woods or paths. It is edible and suitable for soups, but if in doubt one should avoid all grey *Tricholoma* spp. Has a fungal, not mealy, smell. **Cap** 4–8 cm, convex expanding, dry, radially fibrillose, exposing white flesh. **Gills** sinuate, white, often greyish towards margin, crowded. **Stem** 3–8 cm, cylindric, solid, dry, whitish, fibrillose. *Sept–Nov*.

Parks, roadsides and wasteland

## *Macrolepiota rhacodes*
### Shaggy parasol

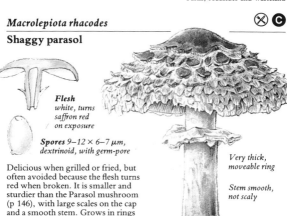

**Flesh** *white, turns saffron red on exposure*

**Spores** *9–12 × 6–7 μm, dextrinoid, with germ-pore*

Delicious when grilled or fried, but often avoided because the flesh turns red when broken. It is smaller and sturdier than the Parasol mushroom (p 146), with large scales on the cap and a smooth stem. Grows in rings on disturbed soil and compost heaps. **Cap** 5–15 cm, dry, hemispherical expanding, grey-brown, with central disc and shaggy margin. **Gills** free, whitish grey, reddening, crowded. **Stem** 10–16 cm, greyish white, dry; ring white, scaly. *Aug–Oct.*

*Very thick, moveable ring*

*Stem smooth, not scaly*

*Large, bulbous stem base*

× 0.3

## *Entoloma sinuatum*

**Spores** *9–12 × 7–9 μm, polyhedral*

× 0.3

*Large, fleshy fruitbodies*

This is a highly poisonous sp, causing severe cramp and vomiting within 30 minutes. Looks like a fleshy *Tricholoma* sp, but has pink spores and is often mistaken for the Clouded agaric (p 13) or the St George's mushroom (p 141). Commonly called *E. lividum*. **Cap** 5–20 cm, convex with an inrolled margin, yellowish grey, shiny, becoming brownish grey, radially fibrillose. **Gills** adnexed, pale yellow often tinted greenish, finally pink, spaced. **Stem** 6–14 cm, hard, solid, silky, white spotted yellowish. **Flesh** firm, white, smell mealy or rancid. *July–Oct.*

## *Lepiota cristata*
### Stinking parasol

**Spores** *6–8 × 2–3 μm*

× 0.6

The thin white flesh has an unpleasant rubbery smell. **Cap** 2–5 cm, convex, umbonate, white, silky, with small red-brown crowded scales, especially near centre. **Gills** free, white, very crowded. **Stem** 2–6 cm, slender, white, with narrow white ring. *Aug–Oct.*

Parks, roadsides and wasteland

## *Agrocybe praecox*
### Spring agaric

*Spores
8–11 ×
5–7 µm,
ellipsoid,
with germ-pore*

## *A. dura*

*Spores
10–14 ×
6–8 µm,
ellipsoid,
with germ-pore*

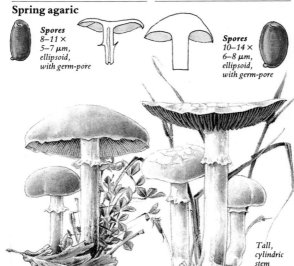

× 0.5

*Tall, cylindric stem*

× 0.5

Found in spring, often by roadsides. Edible, but mediocre, it is recognized by its large thin ring and brown gills. **Cap** 2–6 cm, convex expanding, ochre yellow to pale brown, smooth, with shaggy margin. **Gills** adnexed, pale brown to reddish brown, thin, broad, very crowded. **Stem** 6–9 cm, slender, fibrillose, whitish discolouring brown; ring high, white membranous. *Apr–July.*

Less common than the Spring agaric (left) and more robust with a pale cap that cracks as it dries. This tall mushroom may be very conspicuous on lawns and cultivated land. **Cap** 3–7 cm, convex expanding, white to cream-coloured, dry, smooth, finally cracking. **Gills** adnate, dark brown, crowded. **Stem** 4–10 cm, whitish, with thin ring. *Apr–Aug.*

## *A. erebia*

*White ring finally turns brown*

× 1

*Spores
10–14 ×
7–9 µm,
ellipsoid,
with germ-pore*

Typically found on shady embankments amongst loose soil but only in the autumn. It lacks the mealy smell of the 2 previous spp. **Cap** 2–6 cm, convex-umbonate, dark brown when moist drying to clay colour, smooth, with grooved margin. **Gills** slightly decurrent, pale to cinnamon brown, rather spaced. **Stem** 2–5 cm, grey-brown, fibrillose, with white, narrow, membranous ring. *Sept–Nov.*

Parks, roadsides and wasteland

## *Conocybe tenera*
### Brown cone cap

*Conical cap does not expand*

**Spores** 10–15 × 5–8 µm, ellipsoid, with germ-pore

Usually grows singly, but several of the tall, slender fruitbodies may occur together amongst grass. **Cap** 1–3 cm, conical, ochre brown paling to yellowish, smooth. **Gills** ascending, cinnamon brown, narrow, crowded. **Stem** 8–10 cm, colour as cap, darkening towards base, shiny-striate, finely powdery. *May–Nov.*

× 1

## *Inocybe patouillardii*

**Spores** 9–12 × 5–8 µm, bean-shaped

Deadly: causes muscarine poisoning.
**Cap** 3–7 cm, conical, campanulate, white then yellowish, silky.
**Gills** adnexed, white to olive yellow, then brown, crowded.
**Stem** 4–7 cm, robust, white. *May–Nov.*

*Stains red when handled*

× 0.6

## *Tubaria furfuracea*
### Scurfy tubaria

**Spores** 6–9 × 4–5 µm, ellipsoid

*Scurfy cap*

*Brown, decurrent gills*

Commonly found in large numbers in the autumn, but may occur throughout the year, growing both on the soil and on woody debris. The scurfy cap is much paler when dry. **Cap** 1–4 cm, soon flattened or depressed, wavy, cinnamon brown drying to pale tan, scurfy, with striate margin. **Gills** decurrent, cinnamon brown, rather spaced. **Stem** 2–5 cm, colour as cap, with a white, woolly base. *Jan–Dec.*

*Amongst woody debris*

× 1

135

Parks, roadsides and wasteland

## *Agaricus xanthodermus*
### Yellow-staining mushroom

**Spores**
$5-7 \times 3-4 \,\mu m$, ovoid

*Bruises yellow*

*Large, thick ring*

*Stem base discolours yellow*

One of the few *Agaricus* spp to cause discomfort when eaten; best rejected. The white surface bruises bright chrome yellow, and it has an inky or carbolic smell. In clusters amongst leaf mould. **Cap** 6–15 cm, dry, spherical expanding, white or slightly greyish at centre, with very fine scales. **Gills** free, crowded, pale then flesh pink, finally black-brown. **Stem** 8–15 cm, silky-shiny, white staining yellow. **Flesh** white, chrome yellow in stem base. *July–Oct.*

× 0.5

## *A. bisporus*
### Cultivated mushroom

**Spores**
$6-8 \times 5-6 \,\mu m$, ovoid

*Thick membranous ring*

*Gills blackish brown when mature*

× 0.75

The only commercially grown mushroom in Britain, representing a multi-million pound industry; it is also produced in nearly every other country in the world. The history of its cultivation dates back to 1700. Distinguished from the Field mushroom (p 152) mainly by its large, fleshy ring. The cap of the cultivated form of *A. bisporus* is smooth and pure white, but in the wild it may develop dark brown fibrous scales. Grows in groups on rich soil. **Cap** 5–10 cm, convex then flattened, expanding, white to greyish brown with small, appressed, brown scales. **Gills** free, bright pink to blackish brown, very crowded. **Stem** 3–5 cm, thick, white, with a rather cottony surface; ring white, thick. **Flesh** white, turning slightly red on exposure. *July–Sept.*

Parks, roadsides and wasteland

## *A. bitorquis*

**Spores**
4–7 ×
4–5 µm

*Cap smooth,
not scaly*

Also known as *A. edulis*. Often found in towns, always away from trees, and may appear between paving stones. **Cap** 5–9 cm, hemispherical expanding, white to slightly ochre, yellowing to touch, fibrillose. **Gills** free, pale pink to chocolate brown, very crowded. **Stem** 4–6 cm, solid, with 2 persistent rings; basal ring thin, upper ring more complex. **Flesh** thick, white, firm. *May–Oct.*

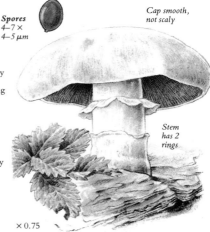

*Stem has 2 rings*

× 0.75

## *A. nivescens*

**Spores**
5–7 ×
4–5 µm,
*ovoid*

*White cap bruises lemon yellow*

The fruitbody is pure white and although it bruises yellow it does not acquire the overall ochre-yellow appearance of the Horse mushroom (p 153). Normally in groups under trees. **Cap** 5–15 cm, strongly convex, expanding, white, silky. **Gills** free, crowded, pale pink to blackish brown. **Stem** 8–10 cm, cylindric, not swollen at base, white discolouring yellow; ring white, thin, broad. **Flesh** thick, smells of almond. *June–Sept.*

*Pink then black gills*

*Smell of almonds*

*Small, white or yellow scales under ring*

× 0.5

Parks, roadsides and wasteland

## *Panaeolina foenisecii*
### Brown hay cap

*Spores 13–17 × 7–9 µm, ellipsoid, warty, with germ-pore*

Sometimes called "Mower's mushroom" because it is very common and numerous on lawns and in short grass from early summer onwards. Has caused mild poisoning in children and there is some evidence that it contains minute amounts of the psychotropic drugs, psilocybin and serotenin. **Cap** 1–2 cm, campanulate to convex, never expanding, dull brown, drying paler from the centre, smooth. **Gills** adnate, mottled, pale brown then umber, broad, crowded. **Stem** 2–5 cm, brown, pale apex, smooth. *June–Oct.*

## *Psilocybe semilanceata*
### Liberty cap

*Spores 11–15 × 7–9 µm, ellipsoid, with germ-pore*

Found in small to large groups but not tufted. It causes psychotropic poisoning, and may produce delirium within a couple of hours. **Cap** 5–10 mm, sharply conical with small, pointed umbo, never expanding, pale tan drying to creamy buff, sticky when wet, with incurved, striate margin. **Gills** ascending, purplish brown with white edge. **Stem** 4–8 cm, tough, wavy, pale. *Aug–Nov.*

## *Panaeolus subbalteatus*

*Spores 11–14 × 7–9 × 6–8 µm, lens-shaped, with germ-pore*

Larger and more robust than other *Panaeolus* spp (p 164), and quite commonly found forming tufts of 2–4 fruitbodies on newly manured soil in gardens. Can cause psychotropic poisoning. **Cap** 2–4 cm, convex expanding, dark brown when moist, drying paler from the centre. **Gills** adnexed, mottled, dark brown, crowded. **Stem** 4–8 cm, cylindric, paler than cap, hollow, striate. *July–Oct.*

Parks, roadsides and wasteland

## *Lacrymaria velutina*
### Weeping widow

*Remains of cortina at cap edge*

**Spores** $9–12 \times 6–7\,\mu m$, lemon-shaped, warty

Common near buildings and roads, often in groups; also found in fields. Deposits of black spores on the dense, cobweb-like cortina give rise to the popular name. Also called *Psathyrella lacrymabunda*. **Cap** 5–10 cm, campanulate to umbonate, fleshy, dull clay brown, fibrillose, with incurved margin. **Gills** adnate, mottled, brown to black, with black droplets on white edge. **Stem** 5–12 cm, clay colour, fibrillose, with white cortinoid ring which turns black. *Apr–Nov*.

× 0.3

## *Psathyrella gracilis*

**Spores** $10–14 \times 6–7\,\mu m$, with germpore

× 0.5

A tall graceful sp with a brittle cap; sometimes found in large numbers, often in long grass. **Cap** 1–3 cm, campanulate expanding, reddish brown drying to pale tan, with a striate margin. **Gills** adnate, greyish black with a pink edge, thin, crowded. **Stem** 3–10 cm, long, slender, smooth, white, with a short, rooting base. *Aug–Nov*.

## *Coprinus atramentarius*
### Common ink cap

**Spores** $8–10 \times 4–5\,\mu m$

× 0.25

Young specimens with white gills are delicious, but when consumed with alcohol they will produce an "antabuse" reaction within 30 min, causing palpitations and vomiting. In clusters on ground. **Cap** 3–6 cm high, ovoid to campanulate, fleshy, grey, radially striate, scaly at centre. **Gills** adnexed, white then black, broad, densely crowded, soon dissolving into a black ink. **Stem** 6–20 cm, white, tapers below, smooth, with small ring near base. *Apr–Nov*.

Parks, roadsides and wasteland

## *Coprinus domesticus*

Gills soon dissolve into black ink

**Spores** 6–10 × 3–5 μm, bean-shaped, with germ-pore

Similar to the Glistening ink cap (p 91), but solitary or in groups of 2 or 3, rather than in large tufts. Usually found on dead wood, often in gardens, but sometimes on old walls or even in damp houses. The veil often leaves a few granular scales on the cap surface. **Cap** 2–4 cm, ovoid then campanulate, date brown, deeply grooved. **Gills** adnexed, white soon black, very crowded, deliquescent. **Stem** 3–7 cm, cylindric, white, silky. *Apr–Dec.*

× 0.5

## *C. comatus*

### Lawyer's wig, Shaggy ink cap

Covered with recurved, fibrous scales

**Spores** 10–14 × 6–8 μm, ellipsoid, with germ-pore

Occasionally entire fields may be covered in these shaggy ink caps, but they are more typically found in small tufts on lawns, compost heaps and roadside verges. Edible with a delicate flavour when young, but they soon spoil. **Cap** 6–15 cm high, cylindric, white then pink or brown, silky-fibrillose, with thick recurved scales, becoming blackened to produce a thick black ink and upturned margin. **Gills** ascending, white then purplish black, very crowded. **Stem** 10–25 cm, white, tapers above; ring thin. *Apr–Nov.*

× 0.3

## *C. plicatilis*

### Little Jap umbrella

**Spores** 10–14 × 9–11 × 5–6 μm, lens-shaped, with germ-pore

Common on lawns and grass verges; often solitary. The thin, pleated cap resembles a Japanese parasol. **Cap** 2–3 cm, soon flat with a depressed centre, light grey or brownish with brown central disc, radially grooved. **Gills** free, narrow, grey to black, spaced, not deliquescent. **Stem** 4–8 cm, white, smooth, fragile. *May–Nov.*

× 0.6

# Heaths and grassland

Much of the grassland in northern Europe owes its formation and continued existence to the activities of humans, for if left undisturbed it would naturally develop into woodland. The removal of trees through burning, grazing and farming, apart from creating a more exposed environment, has resulted in the almost complete absence of many of the mycorrhizal mushrooms, such as *Amanita*, *Cortinarius*, *Lactarius*, *Russula* and the boletes. Instead the more familiar edible mushrooms, including the Field mushroom (p 152), the Horse mushroom (p 153) and the Parasol mushroom (p 146), are more likely to be found. Heaths and moors generally occur where the soil is acidic (but not a deep acid peat), and the ground cover is often composed of small shrubs, notably heather and gorse. These in turn determine the appearance of certain fungi.

## *Calocybe gambosa*

### St George's mushroom

*Spores* 4–7 × 2–4 μm, ellipsoid

Gills very crowded

Cap with thick flesh and inrolled margin

A springtime mushroom, never found in autumn, forming large fairy rings on chalk grassland. It has a smell of damp meal and a good flavour, which is brought out to the full in soups. Do not confuse it with *Entoloma sinuatum* (p 133), *Inocybe patouillardii* (p 135) or *Clitocybe dealbata* (p 143), all of which are poisonous and can grow in similar situations. **Cap** 5–12 cm, hemispherical to convex with an inrolled margin, fleshy, uniformly white, greyish or apricot. **Gills** sinuate, white to cream, narrow, thin, very crowded. **Stem** 3–8 cm, robust, cylindric, short, powdery above, whitish. **Flesh** thick, white, with mealy smell. *Apr–June.*

× 0.5

## *Calocybe carnea*

**Spores** 4–7 × 2–4 μm, ellipsoid

*Crowded gills*

× 0.6

Found amongst short grass in early autumn. The very crowded white gills underneath the flesh-pink cap are distinctive. **Cap** 2–4 cm, convex then flattened, finally depressed, smooth. **Gills** adnate, narrow. **Stem** 2–3 cm, slender, flesh pink, tough. **Flesh** thin, white. *July–Oct.*

## *Camarophyllus pratensis*
### Buff meadow cap

**Spores** 5–7 × 4–6 μm, ovoid

× 0.5

An excellent edible sp, also found in frondose woods. **Cap** 3–9 cm, fleshy, convex, smooth or cracking when dry, buff to tawny yellow. **Gills** deeply decurrent, thick, broad, spaced, pale buff like stem. **Stem** 5–8 cm, colour as cap but paler, tapering below. *Sept–Oct.*

## *Camarophyllus virgineus*

**Spores** 9–12 × 5–6 μm, oblong

*Thick, distant gills*

× 0.5

Similar to *C. niveus* but typically larger and more robust. Common on pastures and downland, sometimes in grassy places in woods. **Cap** 4–7 cm, convex then flattened or depressed, pure white drying yellow. **Gills** decurrent, white, thick, broad, spaced. **Stem** 5–10 cm, white, tapers towards base. *Sept–Oct.*

## *C. niveus*
### Snowy meadow cap

**Spores** 7–12 × 5–6 μm, ellipsoid

*Snow white when dry*

× 0.6

Excellent to eat but small and thin-fleshed. Grows in large troops on short grass. Distinguished from the white *Clitocybe* spp by its thick gills. **Cap** 2–3 cm, convex expanding, finally centrally depressed, smooth. **Gills** short-decurrent, interveined, white, very spaced. **Stem** 3–5 cm, tapers below, soon hollow, white. *Sept–Dec.*

Heaths and grassland

## *Camarophyllus russocoriaceus*

**Spores**
7–9 ×
4–5 µm,
ellipsoid

×1

Small, similar to *C. niveus* (p 142), but the white flesh smells strongly of Russian leather or sandalwood. **Cap** 1–2 cm, convex, slightly sticky, ivory white. **Gills** decurrent, thick, white, spaced. **Stem** 1–4 cm, tapers below, pure white. *Sept–Dec.*

## *Cantharellula umbonata*

**Spores**
8–11 ×
3–4 µm,
spindle-shaped,
amyloid

*Forking gills*

×0.5

Often in large numbers in north. **Cap** 2–4 cm, convex, umbonate then depressed, ash grey with white edge. **Gills** decurrent, forking, white, crowded. **Stem** 5–8 cm, cap colour. *Aug–Nov.*

## *Clitocybe dealbata*

**Spores**
4–5 ×
2–3 µm,
ovoid

*Pinky-brown zones on cap*

×0.5

Do not confuse this dangerous white sp with either Fairy ring champignon (p 131) or the Miller (p 149). The thin flesh contains muscarine, a toxin that produces blurred vision, sweating and twitching within 30 min. Often at edge of woods. **Cap** 2–4 cm, convex expanding, white to beige. **Gills** adnate-decurrent, thin, creamy flesh colour, crowded. **Stem** 2–4 cm, cap colour, fibrillose-silky. *Aug–Nov.*

## *C. rivulosa*

**Spores**
3–5 ×
2–3 µm,
ovoid

*Cap greyish, wrinkled*

×0.5

Distinguished from *C. dealbata* by its wrinkled rather than zoned cap surface. Grows in troops, often forming fairy rings on lawns, and also causes muscarine poisoning. **Cap** 2–5 cm, flat to depressed, or umbonate, at first white then greyish pink to yellowish. **Gills** short-decurrent, paler than cap, crowded. **Stem** 2–5 cm, cap colour, short. **Flesh** white, soft. *Aug–Nov.*

Heaths and grassland

## *Hygrocybe conica*
### Conical slimy cap

**Spores** 7–10 × 4–5 μm, ellipsoid

This brightly coloured sp differs from *H. nigrescens* in the more conical cap and yellow stem base. **Cap** 1–4 cm, acutely conical to umbonate, orange or yellow with reddish tints, finally black. **Gills** free, white to yellow, not red. **Stem** 3–5 cm, cap colour. *July–Nov.*

## *H. calyptraeformis*
### Pink meadow cap

**Spores** 7–8 × 4–5 μm, ellipsoid

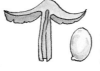

Cap and gills pink or lilac

× 0.3

A fragile mushroom with an easily splitting pointed cap. The lilac or rose-pink tints gradually fade. **Cap** 4–10 cm, acutely conical, expanding, umbonate. **Gills** adnexed, spaced. **Stem** 6–12 cm, tapers both ends, white, splitting. *Aug–Oct.*

## *H. nigrescens*

**Spores** 8–11 × 5–6 μm, ellipsoid

× 0.5

Larger, more robust than *H. conica*; also has a white stem base. This predominantly red fungus finally turns totally black. Mostly found in grass in fields, occasionally in woods. **Cap** 3–8 cm, obtusely umbonate, red, scarlet or yellowish, blackens quickly. **Gills** adnexed, white tinged yellowish green, blackening, very broad, spaced. **Stem** 4–8 cm, lemon yellow, flame red, streaked with black. *Sept–Oct.*

## *H. ceracea*

**Spores** 5–7 × 3–4 μm, ellipsoid

Decurrent gills

All parts of this small, delicate sp are a uniform wax yellow. It can be separated from other yellow spp by the decurrent gills. Grows in large numbers in short grass. **Cap** 1–3 cm, convex, soon flat, sticky, with striate margin. **Gills** triangular, with interveining. **Stem** 2–5 cm, tapers below, hollow. *Sept–Dec.*

Heaths and grassland

## H. chlorophana

**Spores** 6–9 × 4–6 μm, ellipsoid

× 0.5

A slimy sp with consistently lemon-yellow gills. **Cap** 2–4 cm, lemon or chrome yellow, sometimes with an orange flush. **Gills** sinuate-adnate, spaced. **Stem** 4–8 cm, cylindric, smooth, cap colour. *Aug–Oct.*

## H. psittacina
### Parrot toadstool

**Spores** 7–10 × 4–6 μm, ellipsoid

× 0.5

*Cap umbo blue-green*

Covered with a green slime which leaves yellow, red or purple tints. **Cap** 2–5 cm, convex expanding. **Gills** adnate, at first green, thick, quite crowded. **Stem** 4–7 cm, green then yellow, tough, slimy. *July–Nov.*

## H. coccinea
### Scarlet hood

**Spores** 7–9 × 4–5 μm, ellipsoid

*Smooth, dry cap*

× 0.5

Similar in colouring to *H. punicea* (p 146) but smaller in size. The cap and upper part of stem remain blood red. Found in short grass, sometimes in woods. **Cap** 2–5 cm, campanulate, bright scarlet to blood red, fading to orange-buff. **Gills** adnate with decurrent tooth, yellow to red, spaced. **Stem** 4–7 cm, cap colour, smooth. *Aug–Oct.*

## H. miniata

**Spores** 7–10 × 5–6 μm, ellipsoid

*Cap scurfy-scaly at centre*

× 0.5

Prefers damp places. The cap surface is less shiny than in other red spp, with small, scurfy scales at centre. **Cap** 1–2 cm, convex, blood red, scarlet or vermilion, paling to yellow. **Gills** adnate to decurrent, orange with yellow edge. **Stem** 2–5 cm, scarlet, smooth. *June–Oct.*

## *Hygrocybe punicea*

**Spores**
8–12 ×
4–6 μm,
ellipsoid

The largest of the *Hygrocybe* spp and
certainly one of the most beautiful.
Normally the flesh is white, but
the form called *splendidissima* has
yellow flesh (as in section). This
sp usually grows singly in long grass
and is edible, but has a brittle, watery
texture with little taste. These
brightly coloured, shiny *Hygrocybe*
spp are known collectively as wax
agarics. **Cap** 5–12 cm, campanulate,
blood red discolouring yellowish
from the centre, sticky when moist.
**Gills** adnexed, at first yellow tinted
with red, thick, spaced. **Stem** 6–
11 cm, yellow or blood red, white at
base, fibrillose-striate. *Sept–Nov*.

× 0.25

## *Macrolepiota procera*
### Parasol mushroom

**Spores**
14–18 ×
9–12 μm,
ellipsoid,
thick-walled,
dextrinoid,
with a small
germ-pore

Stem surface
with snake-
skin pattern

One of the best edible
spp, with a nutty flavour.
Best to reject fibrous
stems. Young specimens
resemble a drumstick,
but in exceptional cases
the cap can expand to
40 cm diameter. Grows
in grass, generally near
trees, sometimes forming
fairy rings. **Cap** 10–
30 cm, ovoid then convex,
umbonate, with a thick,
shaggy margin, surface
covered with large,
plate-like brown scales
on a greyish brown back-
ground. **Gills** free,
creamy white, broad,
very crowded. **Stem**
15–35 cm, elongate with
a bulbous base, separable
from the cap, brown,
with transverse scaly
bands on a paler back-
ground; ring double,
thick, moveable. *July–Oct*.

× 0.25

Heaths and grassland

## *Macrolepiota excoriata*

*Spores 12–15 × 8–9 µm, dextrinoid*

× 0.3

Occasionally forms large troops, especially on cultivated land and meadows. **Cap** 4–8 cm, conical to convex, umbonate, cream to greyish, with very fine scales. **Gills** free, white, crowded. **Stem** 7–8 cm, white, silky, smooth; ring double, membranous. *May–Nov*.

## *Leucoagaricus naucinus*

*Spores 7–10 × 4–6 µm, dextrinoid*

× 0.5

*White, free gills*

Also known as *Lepiota leucothites*. Resembles an *Agaricus* sp but the gills remain whitish. **Cap** 4–8 cm, convex expanding, fleshy, white or pale brown at centre, smooth, dry, cracking. **Gills** crowded. **Stem** 3–7 cm, white; ring simple. *July–Oct*.

## *Gerronema fibula*
### Carpet-pin mycena

*Spores 4–6 × 2–3 µm, ellipsoid*

*Decurrent gills*

× 1

The var *swartzii* has a brown cap. **Cap** 3–10 cm, convex, flat, pale orange-yellow, striate, translucent. **Gills** white or yellow, broad. **Stem** 2–5 cm, darker below. *Jan–Dec*.

## *Omphalina ericetorum*
### Umbrella navel cap

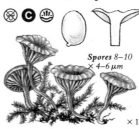

*Spores 8–10 × 4–6 µm*

× 1

On peaty ground. **Cap** 1–2 cm, convex, flat, whitish to olive brown, grooved, with wavy edge. **Gills** triangular, pale yellow, spaced. **Stem** 1–2 cm, cap colour, with woolly base. *Apr–Dec*.

## *O. rustica*

*Spores 6–9 × 3–5 µm, ellipsoid*

× 1

Has narrower gills than *O. ericetorum*. **Cap** 0.5–1 cm, convex, centrally depressed, thin, grey-brown. **Gills** decurrent, grey, spaced. **Stem** 1–2 cm, grey-brown, shiny. *Aug–Oct*.

Heaths and grassland

## *Marasmius graminum*

## *Mycena aetites*

## *M. flavoalba*

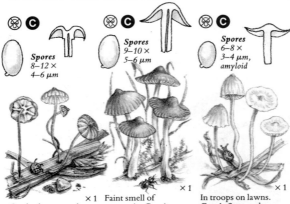

**Spores** 8–12 × 4–6 μm

**Spores** 9–10 × 5–6 μm

**Spores** 6–8 × 3–4 μm, amyloid

×1

On dead stems and leaves of grasses. **Cap** 0.5–0.7 cm, umbonate, red-brown, radially grooved. **Gills** adnate, attached to a collar, cream, very spaced. **Stem** 2–4 cm, deep brown, paler at apex, shiny. *July–Dec.*

×1

Faint smell of ammonia. **Cap** 1–2 cm, dark grey, brownish at centre, campanulate, radially grooved. **Gills** adnate, white to grey, interveined. spaced. **Stem** 3–5 cm, greyish brown. *Aug–Oct.*

×1

In troops on lawns. **Cap** 1–2 cm, pale yellow, darker centre, campanulate expanding, slightly striate. **Gills** adnate with decurrent tooth, white, spaced. **Stem** 2–3 cm, smooth. *Aug–Nov.*

## *Lepista luscina*

## *Melanoleuca grammopodia*

**Spores** 4–6 × 2–4 μm, warty

Mealy smell

**Spores** 6–10 × 4–6 μm, amyloid

× 0.5

Excellent to eat, but do not confuse with *Clitocybe dealbata* (p 143). **Cap** 5–9 cm, convex, finally depressed, pale grey to dark brownish grey, fibrillose, typically spotted. **Gills** short-decurrent, white then grey, narrow, crowded. **Stem** 2–5 cm, solid, grey-brown, fibrillose. **Flesh** thick, whitish. *Oct–Nov.*

× 0.3

In meadows and grassy woodlands. Recognized by the thick, fibrous stem and crowded white gills, which discolour creamy brown. **Cap** 7–15 cm, convex expanding, livid brown to brownish black, smooth, moist. **Gills** sinuate, white then brownish. **Stem** 7–10 cm, white with brown fibrils. *Sept–Nov.*

## Heaths and grassland

### *Crinipellis stipitaria* ⊗ C

**Spores** 7–11 × 5–6 μm, ovoid

### *Clitopilus prunulus* ⊗ C
# The miller

**Spores** 8–13 × 4–7 μm, ridged

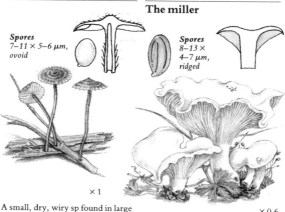

× 1

A small, dry, wiry sp found in large numbers on dead twigs and stems. The fine hairs on the cap and stem are distinctive. **Cap** 0.5–1 cm, convex, umbonate, with long, brown, fibrous, radial hairs on pale ochre-yellow background. **Gills** sinuate, white, narrow, crowded. **Stem** 2–5 cm, brown. *July–Nov*.

× 0.6

Delicious, with a thick, soft flesh and strong mealy smell. Do not confuse with the poisonous white *Clitocybe* spp. **Cap** 3–8 cm, convex to depressed, irregular, dry, whitish, shiny. **Gills** decurrent, whitish to pink, crowded. **Stem** 2–5 cm, often excentric, solid, white. *July–Nov*.

### *Entoloma porphyrophaeum* ⊗ F

**Spores** 9–12 × 6–8 μm, ovoid-angular

× 0.25

### *Leptonia serrulata* ⊗ F

**Spores** 8–10 × 5–7 μm, ovoid-angular

### *L. lampropus* ⊗ F

**Spores** 10–12 × 6–8 μm, ovoid-angular

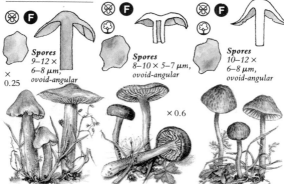

A tall sp, found at edge of woods. **Cap** 4–10 cm, campanulate expanding, umbonate, grey-brown with purplish tints, drying paler, fibrillose. **Gills** adnexed, grey then purplish, spaced. **Stem** 8–13 cm, colour as cap; white base. *Aug–Oct*.

Recognized by the pinkish gills, which have a black edge. **Cap** 1–2 cm, convex with small depression, fibrillose, blackish blue to grey-brown. **Gills** adnate, broad. **Stem** 2–5 cm, blue-grey, paler than cap; base white, woolly. *June–Oct*.

× 0.6

× 0.6

Either solitary or in small groups. **Cap** 1–3 cm, convex finally depressed, dark greyish brown, finely scaly. **Gills** adnate, whitish to pale pink, broad. **Stem** 2–4 cm, slender, smooth, brittle, azure blue. *Aug–Nov*.

## *Leptonia incana* | *L. sericella* | *L. babingtonii*
### | White leptonia |

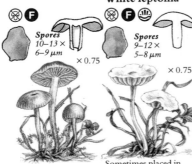

**Spores** 10–13 × 6–9 μm  ×0.75

**Spores** 9–12 × 5–8 μm  ×0.75

**Spores** 11–17 × 6–9 μm  ×1

Strong smell of mice. One of the few green toadstools, but colour varies. **Cap** 1–2 cm, convex to umbilicate, striate margin, yellowish to olive green, streaked with brown. **Gills** adnate to decurrent, pale green then pink. **Stem** 2–5 cm, tough, hard, yellowish green, hollow. *July–Oct.*

Sometimes placed in a separate genus, *Alboleptonia*, owing to its white colour. In fields and open woods. **Cap** 1–2 cm, campanulate finally depressed, white or pale cream, slightly scaly, with incurved margin. **Gills** white then pink, decurrent. **Stem** 2–5 cm, white, translucent, smooth, shiny. *July–Oct.*

A tiny, scaly-capped sp, typically found on damp soil under ferns. The long, angular spores have led some authorities to list it under the genus *Pouzaromyces*. **Cap** 1–2 cm, conical to campanulate, ash grey, shiny, with dark brown, loose fibrils. **Gills** adnate, spaced, grey. **Stem** 2–3 cm, slender, with brown fibrils. *Oct–Nov.*

## *Nolanea sericea*  | *N. papillata*
### Silky nolanea |

**Spores** 7–9 × 6–8 μm, polygonal

**Spores** 8–12 × 6–8 μm, ovoid-angular

×0.5

×0.75

Common in troops in short grass. The cap becomes paler with a silky sheen as it dries and it has a strong mealy smell. **Cap** 2–4 cm, convex expanding, umber brown, striate. **Gills** adnexed, greyish to pink, crowded. **Stem** 2–5 cm, grey-brown, fibrillose. *July–Oct.*

More slender than *N. sericea* (left) with a pointed chestnut-brown cap and lacking a mealy smell. **Cap** 2–3 cm, convex with a papilla, striate. **Gills** adnate, pinkish brown, crowded. **Stem** 2–3 cm, cap colour, shiny. *Sept–Oct.*

Heaths and grassland

## N. staurospora

**Spores**
7–12 ×
6–9 µm,
squarish

× 0.5

Common in fields and open woods; also found in conifer woods and bogs. **Cap** 2–5 cm, campanulate expanding, date brown and striate when moist, drying paler, silky. **Gills** adnexed, pink, crowded. **Stem** 4–8 cm, grey-brown, striate; base white, woolly. *June–Oct*.

## Cortinarius acutus

**Spores**
10–12 ×
5–7 µm,
almond-shaped

× 1

A small, non-sticky *Cortinarius* sp of heaths and conifer woods. Look for a faint white cortina between stem and cap edge. **Cap** 1–2 cm, conical, with pointed umbo, honey-coloured, drying to ochre; margin striate. **Gills** adnate, ochre to cinnamon, fairly crowded. **Stem** 5–7 cm, wavy, cap colour, soon hollow. *Aug–Nov*.

## Agrocybe semiorbicularis

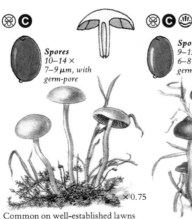

**Spores**
10–14 ×
7–9 µm, with
germ-pore

× 0.75

Common on well-established lawns and near roadsides from early summer onwards. **Cap** 1–2 cm, hemispherical, yellowish to pale tan, smooth. **Gills** pale to deep cinnamon brown, adnate, crowded. **Stem** 6–8 cm, slender, yellowish brown, smooth, shiny. *June–Oct*.

## Hypholoma ericaeum

**Spores**
9–12 ×
6–8 µm, with
germ-pore

× 0.5

A heathland sp with a long, tough stem, found only in damp places. **Cap** 3–4 cm, convex expanding, slightly umbonate, sticky when moist, tawny, smooth. **Gills** adnate, black, crowded. **Stem** 6–10 cm, slender, yellowish above, tawny below, smooth. *May–Oct*.

## *Stropharia coronilla*

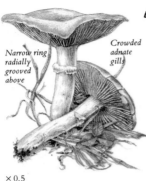

***Spores*** *6–10 × 4–6 µm, ellipsoid, with germ-pore*

Narrow ring radially grooved above

Crowded adnate gills

Commonly grows in groups in fields or at the edge of woods or paths. It is of dubious edibility and easily mistaken for an *Agaricus* sp, but the gills are attached adnately to the stem, not free. **Cap** 2–6 cm, convex expanding, ochre yellow, paler at the margin. **Gills** sinuate-adnate, finally dark brown, crowded. **Stem** 5–8 cm, white to yellowish, slightly scaly below ring; ring white, narrow, striate, attached to upper part of stem. *July–Oct*.

× 0.5

## *Agaricus semotus*

***Spores*** *4–5 × 2–3 µm, ellipsoid*

Old caps become reddish

Found in small groups in meadows, usually under trees, including spruce, but not common. It is similar to *A. purpurellus* (p 35) but the cap has fewer purplish fibrils. **Cap** 3–5 cm, convex, white, with fine, fibrillose, purplish scales at centre. **Gills** free, pale flesh pink, finally blackish, crowded. **Stem** 4–6 cm, cylindric, swollen near base, white bruising yellow; ring thin, white. *July–Oct*.

× 0.5

## *A. campestris*
### Field mushroom

***Spores*** *7–8 × 4–5 µm, ovoid*

Thin, white, fragile ring

The best known wild mushroom yet often confused with Horse mushroom (p 153). The skin does not bruise yellow but the thick white flesh discolours pink. The Cultivated mushroom (p 136) has a larger ring on the stem. The Field mushroom never occurs in woods. **Cap** 3–8 cm, spherical then convex with incurved margin, dry, fibrillose-silky, white. **Gills** free, pale pink then blackish brown. **Stem** 3–5 cm, tapers below, white, firm, silky-scaly; ring simple. *July–Oct*.

× 0.5

## *A. arvensis*
### Horse mushroom

*Spores* 7–9 × 4–5 µm, ovoid

A large, robust species which can occur in great quantities over many years in the same locality, often in circles in fields. It differs from the Field mushroom (p 152) in having a large ring and staining yellow, but do not mistake it for the toxic Yellow-staining mushroom (p 136). The Horse mushroom is noted for its outstanding flavour. **Cap** 8–20 cm, convex expanding, white then yellowing, dry, silky. **Gills** free, crowded, pinkish grey to chocolate brown. **Stem** 6–15 cm, cylindric, fibrillose-silky, white bruising yellow; ring double, white. **Flesh** firm, white, smells of aniseed. *June–Dec.*

White cap discolours yellow

Large, fleshy, pendent ring

× 0.3

Stem base not swollen

## *Melanotus phillipsii*

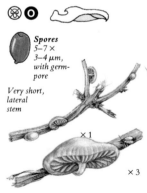

*Spores* 5–7 × 3–4 µm, with germ-pore

Very short, lateral stem

× 1
× 3

A tiny species, easily overlooked, which resembles a small *Crepidotus* but with darker gills. Grows on the dead stems of grasses, herbs and sedges, especially in moist situations. The spore structure relates *Melanotus* to the genus *Psilocybe*. **Cap** 3–8 mm, convex, pale pinkish brown, smooth, striate. **Gills** adnate, pinkish brown, narrow. **Stem** very short, laterally attached (or nearly so), colour as cap. *May–Oct.*

## *Psilocybe physaloides*

*Spores* 6–7 × 4–5 µm

Blackish-brown, adnate gills

× 1

One of several small species of *Psilocybe* which were formerly placed in the genus *Deconica*. Frequently found growing on heaths and moorland, yet may also occur in rich soil near roadsides.
*P. montana* grows in similar situations but has a more striate, darker brown cap. **Cap** 1–1.5 cm, convex expanding, dull bay brown to buff-coloured, striate at edge only when moist, not sticky. **Gills** adnate, pale brown to blackish brown. **Stem** 2–3 cm, slender, paler than cap. *Apr–Oct.*

# Marshes, fens and bogs

Marshy areas occur where the water table is at or near the ground surface and only mushrooms that can tolerate the prevailing acid or alkaline conditions can grow in this habitat. In fens the water is alkaline or neutral, while in raised bogs there is a deep acid peat. Raised bogs are formed by the continuous growth of *Sphagnum* moss upon dead material which has not fully decomposed. Most of the fungi that live in these specialized environments are small and delicate, flourishing in the moist, still atmosphere provided by the mosses, sedges and ericaceous plants. Some, such as *Pholiota myosotis* (p 157), have long stems so that the spores can be released well clear of the vegetation and surface water.

## *Laccaria proxima*    ## *Eccilia sericeonitida*

**Spores** 7–9 × 5–8 μm, spiny

*Powdery gills*

× 0.3

Closely related to the Deceiver (p 66) but confined to marshland; it is usually larger with a longer stem. **Cap** 2–7 cm, convex expanding to depressed, pinkish brown, scurfy-scaly. **Gills** adnate, flesh pink, powdery, spaced. **Stem** 3–10 cm, colour as cap, slender, fibrous. *Sept–Nov*.

**Spores** 7–11 × 5–8 μm, ellipsoid-angular

*Cap soon depressed*

× 0.75

The genus *Eccilia* represents a small group of pink-spored spp with decurrent gills. This sp, formerly known as *E. undata*, is the most commonly found. Always occurs on very moist soil, either in pastures or in woods. **Cap** 1–3 cm, convex to depressed, dark grey-brown becoming paler, silky, faintly zoned. **Gills** decurrent, pinkish brown, narrow. **Stem** 2–3 cm, cap colour or paler. *Sept–Oct*.

Marshes, fens and bogs

## *Omphalina sphagnicola*

## *Dermocybe uliginosa*

Tawny orange cap

Spores 8–16 × 3–5 µm, spindle-shaped

× 0.75

Spores 8–11 × 5–6 µm, warty

× 0.5

Always with *Sphagnum* moss. **Cap** 1–2 cm, infundibuliform, olive brown to sooty brown, striate. **Gills** pale grey-brown, decurrent, narrow. **Stem** 2–4 cm, paler. *June–Sept.*

Found in woodland in boggy places, especially under willow and alder. It grows in groups, often amongst *Sphagnum* moss, on ground that may be submerged for part of the year. Although not recorded as toxic, it could be confused with the deadly *Cortinarius speciosissimus* (p 30). **Cap** 2–5 cm, convex to depressed, bright tawny orange. **Gills** lemon yellow then ochre to rusty, crowded. **Stem** 3–8 cm, cap colour or paler, yellow at apex, fibrillose-striate, with yellow cortina. **Flesh** lemon yellow, smell of radish. *Sept–Oct.*

## *Cortinarius delibutus*

## *C. paleaceus*

Spores 7–10 × 6–8 µm, almost spherical, warty

× 0.25

Spores 8–10 × 4–5 µm, warty

× 0.3

A fairly common sp of the subgenus *Myxacium*, occurring in groups or tufted. Found in wet places in a wide range of habitats, including under conifers, birch and beech. **Cap** 3–9 cm, broadly umbonate, yellow, very sticky. **Gills** adnate, at first deep blue then yellowish to cinnamon brown. **Stem** 5–10 cm, club-shaped, white, with a yellow sticky veil. *Aug–Nov.*

A graceful, slender species of boggy heathland, and sometimes damp woods. It has a distinctive smell of geraniums (*Pelargonium*). **Cap** 1–3 cm, conical with a pointed umbo, expanding, deep brown with whitish recurved fibrillose scales. **Gills** lilac then cinnamon brown, crowded. **Stem** 4–7 cm, wavy, brown with several zones of white scales. **Flesh** brownish lilac. *Sept–Nov.*

## *Galerina sphagnorum*

**Spores** 8–12 × 5–7 μm, almond-shaped, finely warty

×0.3

One of several brown-spored *Galerina* spp with long, slender stems, found amongst *Sphagnum* moss. *G. paludosa* is similar but has a ring on the stem. *G. sphagnorum* smells of meal. **Cap** 1–4 cm, campanulate expanding, umbonate, yellowish brown, striate, smooth. **Gills** adnate, ochre, fairly crowded. **Stem** 4–12 cm, ochre brown, silky. *July–Oct.*

## *G. tibiicystis*

**Spores** 7–12 × 5–6 μm, almond-shaped, warty

*Expanding cap*

×0.3

Like *G. sphagnorum*, this sp is confined to *Sphagnum* moss and lacks a ring on the stem. The stem is finely powdery when viewed under a lens, and there is no mealy smell. **Cap** 2–3 cm, conical, expanding, finally depressed, ochre brown, striate, smooth. **Gills** adnate, rusty brown, fairly crowded. **Stem** 5–12 cm, colour as cap, long, slender. *June–Nov.*

## *G. vittiformis*

*Conical, striate cap*

**Spores** 8–12 × 5–7 μm, finely warty

×0.75

This species is not necessarily confined to marshland, for it will also grow amongst moss in fields (but not *Sphagnum* moss). It is also known as *G. rubiginosa*. A microscope is essential to identify any *Galerina* sp accurately. **Cap** 0.5–1.5 cm, conical to convex, expanding, ochre brown to honey colour, smooth, strongly striate. **Gills** adnate, pale ochre yellow to rust, spaced. **Stem** 2–3 cm, ochre above, red-brown below. *Aug–Oct.*

## *Psilocybe inquilina*

**Spores** 7–10 × 4–6 μm, with germ-pore

×0.5

Grows on dead grass and herb stems. Like many of the *Psilocybe* spp it has a sticky pellicle which can be peeled off, but the gills remain brown and do not acquire a purplish tint. **Cap** 0.5–1 cm, convex expanding, umbonate, dingy brown drying to pale tan, striate when moist. **Gills** broadly adnate, clay to red-brown, rather spaced. **Stem** 1–3 cm, slender, colour as cap, often wavy. *July–Oct.*

Marshes, fens and bogs

## *Pholiota myosotis* | *Hypholoma udum* | *H. elongatum*

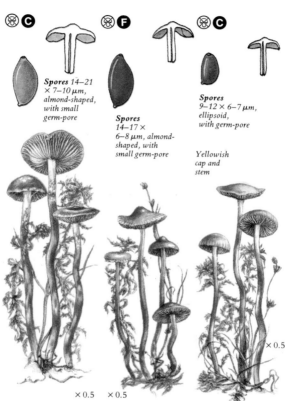

**Spores** 14–21 × 7–10 μm, almond-shaped, with small germ-pore

**Spores** 14–17 × 6–8 μm, almond-shaped, with small germ-pore

**Spores** 9–12 × 6–7 μm, ellipsoid, with germ-pore

Yellowish cap and stem

× 0.5   × 0.5   × 0.5

Frequently found in *Sphagnum* bogs, but probably more common in northern areas. Easily confused with *Hypholoma udum* (right) and microscopic examination of the spores may be necessary to separate the 2 spp. **Cap** 2–4 cm, convex then flat, olive green to light yellowish, darker at centre, smooth, sticky. **Gills** adnate, olive green then rusty brown, fairly spaced. **Stem** 7–15 cm, slender, cylindric, fibrillose, with a white, powdery apex. *July–Sept.*

Differs from *Pholiota myosotis* (left) in having a non-sticky cap, greyish-brown gills and roughened spores. Also it is not unusual to find specimens with a much shorter stem. Grows in peat or with *Sphagnum* moss. **Cap** 1–2 cm, conical to umbonate, yellowish brown, slightly sticky, smooth, veil absent. **Gills** adnate, grey-brown, finally dark brown. **Stem** 4–9 cm, long (esp when young, growing in *Sphagnum*), slender, colour as cap. *Aug–Nov.*

From early summer onwards, large troops of this species may be found pushing up through the *Sphagnum* moss. Recent studies indicate that *H. elongatipes* may be the correct name for this toadstool. **Cap** 1–3 cm, campanulate then flattened, honey yellow with olive tints, and a thin, striate margin. **Gills** adnate, lilac grey, darkening to chestnut brown. **Stem** 5–10 cm, slender, white at apex, yellowish brown below, hollow, smooth. *Sept–Nov.*

# Sand dunes and seaside meadows

The number of species of mushrooms and toadstools which occur on sand dunes is rather small as they have to be specially adapted to contend with a wide temperature range, the continual deposition of sand and the constant shortage of water. Some, such as *Conocybe dunensis* (below) and *Psathyrella ammophila* (p 159), have developed deeply rooting stems in order to seek out any available moisture. A more sheltered habitat is offered in between the dune hills, in the damp hollows called slacks. Here the sparse vegetation provides a more stable layer of soil on which small, delicate mushrooms may flourish.

## *Tephrocybe palustris*

**Spores** 6–8 × 3–5 μm, ellipsoid

× 0.5

Also known as *Collybia leucomyosotis*. Found in bogs, always with *Sphagnum* moss. Smells of meal. **Cap** 1–2 cm, conical or convex, dark brown to greyish brown, striate. **Gills** pale grey, adnexed, spaced. **Stem** 4–11 cm, slender, smooth. *May–Oct*.

## *Omphalina pyxidata*

**Spores** 7–10 × 4–5 μm, ovoid

× 0.5

A small, delicate sp usually found in small troops amongst short grass in sandy areas and near roads. **Cap** 1–2 cm, umbilicate or infundibuliform, orange-brown with dark striations. **Gills** pale yellow, decurrent, spaced. **Stem** 2–3 cm, cylindric or tapering near base, paler than cap, smooth. *Aug–Nov*.

## *Conocybe dunensis*

**Spores** 12–15 × 7–9 μm

Rusty-brown gills

× 0.5

Grows amongst marram grass on sand dunes. **Cap** 1–3 cm, conical or convex, expanding, date brown drying to pale buff, not striate. **Gills** adnate, crowded. **Stem** 4–10 cm, pale ochre, partially buried in sand. *Oct–Nov*.

Sand dunes and seaside meadows

## *Inocybe dulcamara*

**Spores**
$9–12 \times 4–6\,\mu m$,
bean-shaped

× 0.5

In more northerly areas this sp is also found in coniferous woods. Quite common in troops in sandy places. **Cap** 2–5 cm, convex soon flattened, ochre brown, felty-scaly, with shaggy margin when young. **Gills** adnexed, olive yellow finally cinnamon brown, narrow, crowded. **Stem** 4–6 cm, yellow-brown, fibrillose. *Aug–Nov.*

## *Rhodocybe popinalis*

**Spores**
$4–7 \times 4–5\,\mu m$,
ovoid, with irregular ridges

This sp has a strong mealy smell but the bitter taste renders it inedible. Occurs on dunes or in fields.
**Cap** 2–5 cm, convex to umbonate, mouse grey, margin incurved. **Gills** deeply decurrent, greyish or yellowish pink, narrow. **Stem** 1–8 cm, grey. *Aug–Oct.*

## *Psathyrella ammophila*

**Spores**
$10–11 \times 6–7\,\mu m$,
with germ-pore

Grey-brown gills

× 0.5

The stem is always buried deeply in sand, attached to roots of marram grass. **Cap** 1–3 cm, convex with margin incurved at first, date brown then paler. **Gills** pale grey to sepia, spaced. **Stem** 4–5 cm, whitish above, deeply rooting. *Aug–Oct.*

## *Agaricus bernardii*

**Spores**
$5–10 \times 5–6\,\mu m$,
ellipsoid

Deeply cracked cap

Smell foetid or fishy

Found in meadowland washed by sea spray from June onwards. Unlike most *Agaricus* spp it is inedible and has a very unpleasant smell.
**Cap** 10–20 cm, hemispherical expanding with inrolled margin, fleshy, very irregular, white or greyish, cracking into large scales. **Gills** free, pale grey to blackish brown, crowded. **Stem** 5–8 cm, short, hard, solid, whitish, scaly; ring thin. **Flesh** thick, white, reddens on exposure. *June–Oct.*

× 0.3

# Burnt ground

Burnt areas of ground and charred wood logs offer another specialized habitat for a few mushrooms and toadstools. The substrate has been sterilized by fire so there is little competition from other plant life. Some fungal spores are known to require stimulation by heat before they will germinate and grow into a branching network of hyphal threads from which the fruitbodies are formed. Most of the mushrooms are rather dull coloured and may easily be overlooked on the dark background.

## *Tephrocybe anthracophila* | *T. atrata*

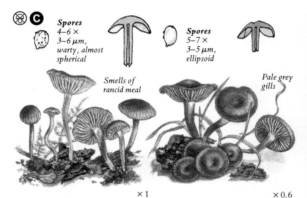

**Spores** 4–6 × 3–6 μm, warty, almost spherical

Smells of rancid meal

× 1

**Spores** 5–7 × 3–5 μm, ellipsoid

Pale grey gills

× 0.6

More widely known as *Collybia carbonaria*, and often erroneously named *C. ambusta* in earlier books. Usually grows in small clusters on burnt soil. **Cap** 1–2 cm, convex expanding, finally flattened, blackish brown, with a striate margin. **Gills** adnate, white then greyish, spaced. **Stem** 2–3 cm, slender, wavy, colour as cap except for a paler apex. *Aug–Nov*.

Found on charcoal heaps and burnt soil. This species also has a mealy smell, similar to *T. anthracophila* (left), and the two species may only be reliably separated by the spore characters under the microscope. **Cap** 2–4 cm, slightly depressed, black when moist, drying a paler sooty brown, smooth. **Gills** adnate, white then grey, spaced. **Stem** 2–5 cm, sooty brown, tough, smooth. **Flesh** brown, thin. *July–Dec*.

Burnt ground

## *Myxomphalia maura*

⊗ **F** ⚠

**Spores** 5–6 × 3–5 μm, ovoid, amyloid

*Cap deeply depressed*

× 0.5

Always on burnt ground, mainly in coniferous woods, and often in great numbers. Widespread but not very common. **Cap** 2–4 cm, deeply umbilicate, dark grey-brown drying paler. **Gills** decurrent, white, very crowded. **Stem** 2–5 cm, rigid, grey-brown to black, smooth. **Flesh** greyish, with fruity smell. *Sept–Nov.*

## *Hebeloma anthracophilum*

⊗ **C**

**Spores** 10–13 × 5–7 μm, almond-shaped, warty

*Sticky cap with tawny centre*

◀ *Brown sinuate gills*

× 0.5

Confined to burnt sites, often those which have been colonized by moss. The tough, elastic flesh has a pleasant smell but bitter taste. **Cap** 3–5 cm, hemispherical expanding, tawny brown centre, paler elsewhere. **Gills** sinuate, clay to rusty brown, crowded. **Stem** 5–8 cm, cylindric, white, silky, becoming brown from the base up. *Sept–Nov.*

## *Pholiota highlandensis*
### Charcoal pholiota

⊗ **C**

**Spores** 6–8 × 4–5 μm, ellipsoid

*Smooth, sticky cap*

× 0.3

Better known as *P. carbonaria*. Forms small clusters on burnt stumps and fire sites. **Cap** 2–7 cm, convex, flattened, smooth, sticky, chestnut brown then pale; margin incurved. **Gills** adnate, clay then dull brown, crowded. **Stem** 2–7 cm, scaly-fibrillose, yellowish to brownish. *Sept–Nov.*

## *Psathyrella pennata*

⊗ **C**

**Spores** 7–8 × 3–4 μm, with small germ-pore

× 0.5

These delicate, brittle fruitbodies occur in small troops on scorched soil; recognized by the loose white scales covering cap and stem. **Cap** 1–3 cm, campanulate expanding, ochre yellow, at first covered in white fibrils. **Gills** adnate, white then brownish black. **Stem** 3–5 cm, white. *Sept–Nov.*

# Dung

Fungi growing on dung are described as coprophilous. Their occurrence is related to the distribution of herbivores, such as cattle, horses, sheep, rabbits and deer. Released spores which have fallen on to the vegetation are eaten by the animal, pass through the gut and are finally excreted. Animal dung provides a substratum extremely rich in nitrogenous material, which has been largely sterilized by the high temperature and digestive enzymes of the gut. The spores are thick-walled and remain viable so they can germinate in the dung without competition from other organisms.

## *Bolbitius vitellinus*

### Yellow cow-pat toadstool

**Spores**
$12–15 \times 7–9\,\mu m$, ellipsoid, with germ-pore

Mostly amongst grass on horse or cow dung. A delicate sp with a sticky flat cap and hollow stem. **Cap** 2–4 cm, campanulate soon flat, chrome yellow at centre then paler, very thin, with striate margin, splitting. **Gills** adnexed, brown, crowded, thin. **Stem** 6–10 cm, slender, striate, whitish or yellowish, powdery. *Aug–Nov*.

## *Conocybe pubescens*

**Spores**
$12–18 \times 7–10\,\mu m$, ellipsoid, with germ-pore

Probably the most common of the several *Conocybe* spp found on dung. It is usually smaller than the related *C. rickenii* (p 163). **Cap** 1–3 cm, convex to campanulate, tawny brown, drying paler, striate when moist. **Gills** adnate, rust brown, crowded. **Stem** 4–7 cm, slender, wavy, white above, brown below, striate. *Aug–Oct*.

Dung

## Conocybe rickenii

**Spores** 13–18 × 8–9 μm, ellipsoid, with germ-pore

× 0.3

Often found in enormous numbers in gardens where the soil has been enriched with horse manure. **Cap** 1–4 cm, conical to umbonate, pale ochre yellow to cream, striate. **Gills** adnate, yellowish brown, spaced. **Stem** 2–10 cm, white to pale brown, powdery. *Aug–Oct*.

## Coprinus niveus

**Spores** 12–18 × 10–13 μm, with germ-pore

× 0.25

One of the larger coprophilous ink-caps, forming small colonies on cow and horse dung. **Cap** 2–4 cm, ovoid to campanulate with margin finally upturned, white, with dense mealy covering. **Gills** soon black, deliquescent, crowded. **Stem** 4–8 cm, tapers above, white, with woolly scales. *May–Nov*.

## Coprinus narcoticus

**Spores** 10–12 × 5–6 μm, with germ-pore and loose perisporium

*Small, flaky scales on cap*

× 0.5

Differs from *C. niveus* (above) by its greyish colour and strong, unpleasant smell of coal gas. The stem sometimes has a deeply rooting base. **Cap** 1–2 cm, conical, almost cylindric, white or greyish, with loose, white, recurved scales. **Gills** free, white then black, narrow. **Stem** 4–5 cm, white, fibrillose. *Oct–Nov*.

## C. patouillardii

**Spores** 5–10 × 4–7 μm, lens-shaped, with germ-pore

*Grooved cap surface*

× 1

Grows on old dung, rotting straw, silage and even spent tea-leaves. It is one of the smaller ink-caps, and has a distinctive grooved cap with a yellowish centre. **Cap** 0.5–1.5 cm, conical to campanulate, expanding, ash grey with ochre-yellow scales at centre, furrowed. **Gills** free, soon black, spaced. **Stem** 1–4 cm, white, smooth. **Flesh** very thin. *July–Oct*.

## Coprinus radiatus

*Cap with white, curved scales*

*Spores 11–14 × 6–7 µm, bean-shaped, with germ-pore*

*Rooted in horse dung* × 1

Although widespread, this tiny inkcap is almost restricted to horse dung and therefore its occurrence depends on the distribution of the horse. **Cap** 4–8 mm, ovoid, expanding, grey, scaly, grooved. **Gills** free, few, soon black, spaced. **Stem** 5–25 cm, white, hairy, with rooting base. *May–Nov.*

## Panaeolus campanulatus

**Spores** 12–14 × 8–9 µm, with germ-pore

*Red-brown cap*

× 0.5

Distinguished by its red-brown cap. It contains ibotenic acid which can cause nausea. Confined to horse dung. **Cap** 2–3 cm, hemispherical to convex, drying yellow-brown, shiny, often cracked. **Gills** black, white-edged, crowded. **Stem** 7–10 cm, whitish. *Aug–Oct.*

## P. sphinctrinus
### Grey mottle gill

**Spores** 14–18 × 9–10 µm, with germ-pore

*Dark grey cap*

× 0.5

On all dung and rich soil. Recognized by the grey-brown to almost black stem and fringe of white scales on the cap edge. **Cap** 2–3 cm, campanulate, not expanding, dark grey. **Gills** adnate, black, white-edged, broad. **Stem** 7–12 cm, erect, powdery. *July–Oct.*

## P. semiovatus
### Dung mottle gill

**Spores** 16–22 × 9–15 µm, with germ-pore

× 0.5

Formerly known as *Anellaria separata*. **Cap** 2–6 cm, whitish to pale brown, sticky when wet, smooth, shiny, campanulate, not expanding. **Gills** adnate, grey to black, mottled, crowded. **Stem** 7–15 cm, rigid, white or yellowish, striate; ring thin, white. *July–Nov.*

Dung

## *Psilocybe coprophila*

**Spores**
12–15 ×
8–10 μm,
almost hexagonal,
with germ-pore

Gills adnate,
blackish

×0.5

Grows in small clusters on old dung and enriched soil. It is frequently confused with *P. merdaria* (below) but the cap is generally smaller and the stem, apart from being more slender, is not scaly, nor does it have a ring. **Cap** 1–2 cm, hemispherical, not expanding, light to dark brown, with a sticky pellicle (easily peeled off), smooth and shiny. **Gills** broadly adnate, greyish brown to black, with a white edge. **Stem** 2–4 cm, paler than cap, with a few white fibrils but no scales. *Aug–Nov*.

## *P. merdaria*

**Spores**
10–13 ×
6–8 μm,
ellipsoid,
with germ-pore

×0.5

This species is sometimes included in the genus *Stropharia* because the partial veil leaves a small ring on the stem, but all other features are typical of the *Psilocybe* group. Small white flaky scales may also be found on the lower stem and cap margin. **Cap** 2–4 cm, rounded expanding to flat, cinnamon brown when moist, drying yellowish to olive green, slightly sticky, smooth. **Gills** adnate, blackish to dark brown, quite crowded. **Stem** 2–5 cm, yellowish brown, with fine white scales below ephemeral ring. *Aug–Oct*.

## *Stropharia semiglobata*

### Dung roundhead

**Spores**
16–24 ×
8–11 μm,
ellipsoid,
with germ-pore

×0.5

A tall, slender fungus, extremely common on dung in fields through most of the year. Note the smooth, rounded cap and the long, narrow stem with a ring zone which becomes coloured black as the spores are released. **Cap** 1–4 cm, pale yellow, sticky. **Gills** adnate, broad, blackish, crowded. **Stem** 8–10 cm, whitish or yellowish, hollow, smooth, sticky, with a thin ring forming a dark zone on the upper part. *July–Nov*.

# Index

Entries in parentheses are mushrooms mentioned in the text but not described in full.

*Agaricus abruptibulbus* 37
  *arvensis* 153
  *augustus* 59
  *bernardii* 159
  *bisporus* 136
  *bitorquis* 137
  *campestris* 152
  *comtulus* 35
  *edulis* 137
  *langei* 36
  *macrosporus* 59
  *meleagris* 19
  *nivescens* 137
  *placomyces* 91
  *purpurellus* 35
  *semotus* 152
  *silvaticus* 36
  *silvicola* 35
  *xanthodermus* 136
*Agrocybe aegerita* 89
  *cylindracea* 89
  *dura* 134
  *erebia* 134
  *praecox* 134
  *semiorbicularis* 151
(*Alboleptonia*) 150
*Amanita citrina* 104
  *excelsa* 44
  *fulva* 96
  *inaurata* 43
  *muscaria* 95
  *pantherina* 44
  *phalloides* 104
  *porphyria* 37
  *rubescens* 44
  *solitaria* 62
  *spissa* 44
  *strangulata* 43
  *vaginata* 96
  *virosa* 104
Amethyst deceiver 66
*Anellaria separata* 164
*Armillaria mellea* 45
  *tabescens* 45

*Baeospora myosura* 19
Bare-toothed russula 76
Bay boletus 39
Bitter boletus 124
Blackening russula 52
Blackish-purple russula 113
Bleeding mycena 67
Blewit 132
Blood-red cortinarius 33
Blue-green clitocybe 63
Blusher 44
Blushing wax agaric 13
*Bolbitius vitellinus* 57
*Boletus aestivalis* 123
  *appendiculatus* 121
  *calopus* 60
  *edulis* 123
  *elegans* 41
  *erythropus* 60
  *luridus* 122
  *reticulatus* 123
  *satanas* 122
  *viscidus* 41
Bonnet mycena 68
Branched oyster fungus 110
Brick-red hypholoma 58
Broad-gilled agaric 17
Brown birch boletus 102
Brown cone cap 135
Brown hay cap 138
Brown roll-rim 101
Buff meadow cap 142

*Calocybe carnea* 142
  *gambosa* 141
*Camarophyllus niveus* 142
  *pratensis* 142
  *russocoriaceus* 143
  *virgineus* 142
*Cantharellula umbonata* 143
*Cantharellus cibarius* 103
  *infundibuliformis* 105
  *lutescens* 66
Carpet-pin mycena 147
Cep 123
*Chalciporus piperatus* 61
*Chamaemyces fracidus* 72
Chanterelle 103
Charcoal pholiota 161
Cinnamon cortinarius 33
*Chroogomphus rutilus* 42
*Clitocybe brumalis* 46
  *candicans* 46
  *clavipes* 14
  *dealbata* 143
  *diatreta* 14
  *dicolor* 64
  *flaccida* 14
  *fragrans* 63
  *geotropa* 64
  *gibba* 45
  *infundibuliformis* 45
  *inornata* 64
  *langei* 14
  *nebularis* 13
  *odora* 63
  *rivulosa* 143
  *suaveolens* 63
*Clitopilus hobsonii* 82
  *prunulus* 149
Clouded clitocybe 13
Club-footed clitocybe 14
Clustered tough shank 105
Coconut-scented milk cap 98
*Collybia* (*acervata*) 47
  *ambusta* 160
  *bresadolae* 47
  *butyracea* 46
  *carbonaria* 160
  *cirrhata* 67
  *confluens* 105
  *dryophila* 106
  *erythropus* 47
  *fusipes* 106
  *gangraenosa* 21
  *leucomyosotis* 158
  *maculata* 15
  *marasmioides* 47
  *peronata* 66
  *tuberosa* 47
Common funnel cap 45
Common ink cap 139
Common white inocybe 84
Common yellow russula 51
Conical slimy cap 144
*Conocybe dunensis* 158
  *pubescens* 164
  *rickenii* 163
  *tenera* 157
*Coprinus atramentarius* 139
  *comatus* 92
  *disseminatus* 92
  *domesticus* 140
  *lagopus* 92
  *micaceus* 91
  *narcoticus* 163
  *niveus* 163
  *patouillardii* 163
  *picaceus* 120

*Coprinus* (*continued*)
  *plicatilis* 92
  *radiatus* 164
*Cortinarius* subg *Hydrocybe* 32
  *Myxacium* 29
  *Phlegmacium* 30
  *Sericeocybe* 100
  *Telamonia* 32
  *C. acutus* 151
  *alboviolaceus* 56
  *anomalus* 118
  *armillatus* 101
  (*azureus*) 57
  *caerulescens* 56
  *caesiocyaneus* 117
  *callisteus* 31
  *collinitus* 29
  (*crocolitus*) 100
  *cyanites* 57
  *decipiens* 32
  *delibutus* 155
  *elatior* 117
  *gentilis* 32
  *glandicolor* 32
  *hemitrichus* 101
  *hinnuleus* 116
  *largus* 83
  *melliolens* 118
  *mucosus* 29
  *ochroleucus* 83
  *paleaceus* 155
  *pholideus* 100
  *pseudosalor* 117
  *purpurascens* 30
  (*saniosus*) 30
  *speciosissimus* 30
  *tabularis* 57
  *torvus* 118
  *traganus* 31
  *triumphans* 100
  *trivialis* 83
*Craterellus cornucopioides* 105
*Crepidotus mollis* 82
  *variabilis* 82
*Crinipellis stipitaria* 149
Cultivated mushroom 136
Curry-scented milk cap 24
*Cystoderma amianthinum* 22
  *carcharias* 22
  *granulosum* 22
*Cystolepiota aspera* 74

Death cap 104
Deceiver 66
*Deconica* 153
*Dermocybe* (*cinnabarina*) 33
  *cinnamomea* 33
  (*cinnamomeo-badius*) 33
  (*cinnamomeo-lutescens*) 33
  *sanguinea* 33
  *semisanguinea* 33
  *uliginosa* 155
Destroying angel 104
Devil's boletus 122
Dung mottle gill 164
Dung roundhead 165

*Eccilia sericeonitida* 154
  *undata* 154
Emetic russula 25
*Entoloma aprile* 80
  *clypeatum* 80
  *lividum* 133

Entoloma (continued)
 nidorosum 80
 nitidum 27
 porphyrophaeum 149
 rhodopolium 80
 sinuatum 133

Fairy cake hebeloma 87
Fairy ring champignon 131
False chanterelle 17
False death cap 104
Fawn pluteus 81
Fayodia bisporigera 49
Field mushroom 152
Flammulina velutipes 67
Fleecy milk cap 77
Fly agaric 95
Foetid russula 52
Fragile russula 53
Fringed crumble cap 92

Galerina mniophila 86
 (paludosa) 156
 rubiginosa 156
 sphagnorum 156
 tibiicystis 156
 unicolor 28
 vittiformis 156
Geranium-scented russula 113
Gerronema chrysophyllum 69
 fibula 147
Giant clitocybe 131
Glistening ink cap 91
Goblet 65
Gomphidius glutinosus 42
 roseus 42
 viscidus 42
Greasy tough shank 46
Grey milk cap 98
Grey mottle gill 164
Grisette 96
Gymnopilus junonius 88
 penetrans 28
Gyrodon lividus 126
Gyroporus castaneus 124
 cyanescens 124

Hebeloma anthracophilum 161
 crustuliniforme 87
 longicaudum 87
 mesophaeum 28
 radicosum 119
 sacchariolens 87
 sinapizans 86
Herald of the winter 13
Hohenbuehelia petaloides 79
Honey fungus 45
Horn of plenty 105
Horse-hair fungus 18
Horse mushroom 153
Hygrocybe calyptraeformis 144
 ceracea 144
 chlorophana 145
 coccinea 145
 conica 144
 miniata 145
 nigrescens 144
 psittacina 145
 punicea 146
Hygrophoropsis aurantiaca 17
Hygrophorus chrysaspis 108
 dichrous 63
 eburneus 108
 erubescens 13
 hypothejus 13
Hypholoma capnoides 38
 dispersum 38
 elongatipes 157
 elongatum 157
 ericaeum 151
 fasciculare 58
 marginatum 38

Hypholoma (continued)
 sublateritium 58
 udum 157

Inocybe asterospora 85
 calospora 86
 cookei 85
 dulcamara 159
 eutheles 84
 fastigiata 119
 flocculosa 85
 geophylla 84
  var lilacina 84
 godeyi 85
 griseolilacina 84
 lacera 57
 lanuginosa 85
 maculata 119
 patouillardii 135
 pyriodora 84

Kuehneromyces mutabilis 89

Laccaria amethystea 66
 laccata 66
 proxima 154
Lacrymaria velutina 139
Lactarius blennius 115
 camphoratus 24
 deliciosus 23
 deterrimus 23
 flexuosus 78
 fuliginosus 55
 glyciosmus 98
 helvus 24
 mitissimus 55
 necator 97
 obscuratus 125
 pallidus 115
 piperatus 77
 plumbeus 97
 pyrogalus 78
 quietus 115
 rufus 23
 scrobiculatus 23
 serifluus 55
 subdulcis 78
 tabidus 98
 torminosus 54
 trivialis 54
 turpis 97
 uvidus 98
 vellereus 77
 vietus 98
 volemus 114
Larch boletus 41
Lawyer's wig 140
Leccinum (aurantiacum) 102
 (holopus) 102
 scabrum 102
 testaceoscabrum 102
 versipelle 102
 (vulpinum) 102
Lentinellus cochleatus 79
Lentinus lepideus 20
 tigrinus 20
Lepiota bucknallii 74
 castanea 73
 clypeolaria 74
 cristata 133
 friesii 74
 fulvella 73
 haematosperma 74
 hetieri 73
 irrorata 72
 leucothites 147
 seminuda 73
 serena 73
 sistrata 73
 (ventriosospora) 50
Lepista irina 65
 luscina 148
 nuda 51

Lepista (continued)
 saeva 132
 sordida 132
Leptonia babingtonii 150
 incana 150
 lampropus 149
 sericella 150
 serrulata 149
Leucoagaricus naucinus 147
Leucopaxillus giganteus 131
Liberty cap 138
Lilac inocybe 84
Lilac mycena 108
Limacella guttata 72
 lenticularis 72
Little Jap umbrella 140
Little wheel toadstool 69
Lyophyllum connatum 129
 decastes 130
 fumatofoetens 21
 fumosum 130
 ulmarium 127

Macrocystidia cucumis 20
Macrolepiota excoriata 147
 procera 146
 rhacodes 133
Magpie ink cap 120
Marasmiellus ramealis 69
Marasmius alliaceus 109
 androsaceus 18
 cohaerens 109
 epiphyllus 69
 globularis 109
 graminum 147
 oreades 131
 rotula 69
 wynnei 109
Marvellous tricholoma 16
Melanoleuca brevipes 130
 grammopodia 148
 melaleuca 75
Melanophyllum echinatum 74
Melanotus phillipsii 153
Micromphale foetidum 49
 perforans 18
Milk-drop mycena 50
Milk-white russula 52
Miller 149
Mower's mushroom 138
Mycena aetites 148
 corticola 68
 crocata 68
 epipterygia 18
 flavoalba 148
 galericulata 68
 galopus 50
 haematopus 67
 inclinata 107
 leptocephala 18
 metata 50
 olida 68
 pelianthina 108
 polygramma 68
 pura 108
 rorida 50
 sanguinolenta 18
 tenerrima 67
 tortuosa 125
 vitilis 107
Myxomphalia maura 161

Naucoria escharoides 126
 scolecina 126
Nitrous mycena 18
Nolanea cetrata 27
 cuneata 27
 papillata 150
 sericea 150
 staurospora 151
Nyctalis asterophora 47
 parasitica 47

Oak milk cap 115
Old man of the woods 94
Omphalina ericetorum 147
  pyxidata 158
  rustica 147
  sphagnicola 155
Orange birch boletus 102
Orange pholiota 88
Oudemansiella mucida 106
  radicata 107
Oyster mushroom 110

Panaeolina foenisecii 138
Panaeolus campanulatus 164
  semiovatus 164
  sphinctrinus 164
  subbalteatus 138
Panther 44
Panus conchatus 71
  stipticus 71
  torulosus 71
Parasol mushroom 146
Paxillus atrotomentosus 39
  involutus 101
  panuoides 39
Pear-scented inocybe 84
Penny bun boletus 123
Penny top 106
Peppery boletus 61
Peppery milk cap 77
Phaeocollybia festiva 28
Phaeolepiota aurea 88
Pholiota adiposa 121
  alnicola 126
  aurivella 90
  carbonaria 161
  flammans 29
  gummosa 89
  highlandensis 161
  lenta 58
  myosotis 157
  spectabilis 88
  squarrosa 90
Phylloporus rhodoxanthus 116
Phyllotopsis nidulans 21
Pick-a-back toadstool 47
Pine spike-cap 42
Pleurotus cornucopiae 110
  dryinus 71
  ostreatus 110
Pluteus atricapillus 81
  cervinus 81
  cinereofuscus 81
  lutescens 81
  salicinus 81
Poison pie 87
(Pouzaromyces) 150
Psathyrella ammophila 159
  candolleana 92
  caput-medusae 34
  conopilea 139
  gracilis 139
  hydrophila 93
  lacrymabunda 139
  obtusata 120
  pennata 161
  spadicea 94
  spadiceogrisea 93
  subatrata 93
Pseudoclitocybe cyathiformis 65
  expallens 65
Psilocybe coprophila 165
  inquilina 156
  merdaria 165
  (montana) 153
  physaloides 153
  semilanceata 138

Psilocybe (continued)
  squamosa 120
Purple and yellow agaric 17

Red-banded cortinarius 101
Red-cracked boletus 61
Resupinatus applicatus 69
Rhodocybe popinalis 159
  truncata 20
Rhodotus palmatus 128
Ripartites tricholoma 19
Roman-shield entoloma 80
Rooting shank 107
Rose-gilled grisette 82
Rough stalks 102
Rozites caperata 34
Rufous milk cap 24
Russet shank 106
Russula adusta 25
  atropurpurea 113
  aurata 112
  azurea 26
  betularum 99
  caerulea 27
  claroflava 99
  consobrina 25
  cyanoxantha 75
  delica 52
  (densifolia) 25
  emetica 25
  fellea 113
  foetens 52
  fragilis 25
  heterophylla 76
  (integra) 76
  (laurocerasi) 52
  lepida 112
  (lutea) 51
  mairei 112
  nigricans 25
  nitida 99
  ochroleuca 51
  paludosa 26
  parazurea 113
  pseudointegra 114
  puellaris 53
  queletii 26
  (sororia) 25
  versicolor 99
  vesca 76
  virescens 114
  xerampelina 53

Saffron milk cap 23
Saffron parasol 22
Scarlet hood 145
Scented hebeloma 87
Schizophyllum commune 110
Scurfy cortinarius 101
Scurfy tubaria 135
Shaggy ink cap 140
Shaggy parasol 155
Shaggy pholiota 90
Sheathed cortinarius 118
Silky nolanea 150
Silky volvaria 128
Slimy beech caps 106
Slimy milk cap 115
Slimy spike-cap 42
Small bleeding mycena 18
Snow meadow cap 142
Soap-scented tricholoma 70
Soft slipper toadstool 82
Spindle shank 106
Split gill 110
Spotted tough shank 15
Spring agaric 134
Steely-stemmed mycena 68
St George's mushroom 141
Stinking parasol 133
Strobilomyces floccopus 94
Strobilurus esculentus 19
  (tenacellus) 19

Stropharia aeruginosa 91
  coronilla 152
  hornemannii 38
  semiglobata 165
Styptic fungus 71
Suillus aeruginascens 41
  bovinus 40
  granulatus 40
  grevillei 41
  luteus 41
  variegatus 40
Sulphur tuft 58
Sulphurous tricholoma 70

Tawny funnel cap 14
Tawny grisette 96
Tephrocybe anthracophila 160
  atrata 160
  palustris 158
Tricholoma aggregatum 130
  albobrunneum 15
  album 97
  argyraceum 111
  cinerascens 130
  columbetta 111
  flavobrunneum 97
  fulvum 97
  imbricatum 16
  leucophaeatum 21
  portentosum 16
  saponaceum 70
  scalpturatum 111
  sejunctum 48
  sulphureum 70
  terreum 132
  ustale 70
  (ustaloides) 70
  (vaccinum) 16
  virgatum 48
Tricholomopsis decora 48
  platyphylla 17
  rutilans 17
Trooping crumble cap 92
Tubaria furfuracea 135
Two-toned pholiota 89
Tylopilus felleus 124

Ugly milk cap 97
Umbrella navel cap 147

Velvet shank 67
Verdigris toadstool 91
Volvariella bombycina 128
  speciosa 82

Watery milk cap 55
Weeping widow 139
White leptonia 91
Willow pluteus 81
Winter funnel cap 96
Wood blewit 51
Wood mushroom 35
Wood woolly-foot 66
Woolly milk cap 54

Xerocomus badius 39
  chrysenteron 61
  subtomentosus 61
Xeromphalina campanella 19
  cauticinalis 49

Yellow cow-pat toadstool 162
Yellow-cracked boletus 61
Yellow-staining mushroom 136
Yellow-stemmed mycena 18
Yellow swamp russula 99